Foodborne Pathogens: Detection Methods, Food Safety, and Public Health

Foodborne Pathogens: Detection Methods, Food Safety, and Public Health

Editor

Abiodun Adewale Adesiyun

MDPI

Basel • Beijing • Wuhan • Barcelona • Belgrade • Novi Sad • Cluj • Manchester

Editor
Abiodun Adewale Adesiyun
University of Pretoria
Pretoria
South Africa

Editorial Office
MDPI
St. Alban-Anlage 66
4052 Basel, Switzerland

This is a reprint of articles from the Special Issue published online in the open access journal *Microorganisms* (ISSN 2076-2607) (available at: https://www.mdpi.com/journal/microorganisms/special_issues/foodborne_pathogens_detection_safety).

For citation purposes, cite each article independently as indicated on the article page online and as indicated below:

Lastname, A.A.; Lastname, B.B. Article Title. *Journal Name* **Year**, *Volume Number*, Page Range.

ISBN 978-3-0365-8896-4 (Hbk)
ISBN 978-3-0365-8897-1 (PDF)
doi.org/10.3390/books978-3-0365-8897-1

Contents

About the Editor

Abiodun Adewale Adesiyun

Abiodun Adewale Adesiyun received his Doctor of Veterinary Medicine (DVM) degree from Ahmadu Bello University, Nigeria. He attended the University of Minnesota, USA, where he completed his Master of Public Health and Ph.D. degrees. He has also served as a Research Fellow at the University of Bonn, Germany. Professor Adesiyun was appointed Senior Lecturer at The University of the West Indies, Trinidad and Tobago, in 1990, and was promoted to the rank of professor in 1999. In the Faculty of Medical Sciences, he has served as the Associate Dean for Research, Deputy Dean of Basic Health Sciences, and acting Dean on several occasions; he also held the position of Director of the School of Veterinary Medicine for eight years (2006-2014). He has been an Emeritus professor at the University of the West Indies since 2016 and recently was deemed an extra-ordinary professor at the University of Pretoria, South Africa (2014-2023). He has taught undergraduate and postgraduate students and supervised over 50 graduate (M.Sc./M.Phil. and Ph.D.) students in Africa (Nigeria and South Africa), the Caribbean (Trinidad and Tobago), and in collaborative research in the USA (Tuskegee University). His areas of interest include the application of molecular methods and epidemiological principles to bacterial pathogens (Staphylococcus aureus, *Salmonella* spp., *Listeria monocytogenes*, Shiga-toxin producing *Escherichia coli* (STEC), *Campylobacter jejuni*) of significance in food safety. He has also published extensively on other bacterial pathogens (*Leptospira* spp. and *Brucella* spp.). Overall, he has published over 250 papers in peer-reviewed journals.

Preface

Standard microbiological methods have, for decades, been employed to isolate, identify, and characterize foodborne pathogens. However, the emergence of molecular methods has revolutionized the detection of both live and dead pathogens in foods, and more importantly, the characterization of pathogens and the provision of invaluable data regarding their pathogenesis and genetic relatedness. In this Special Issue, combinations of phenotypic and molecular approaches are employed. Nine of the ten papers published in this Special Issue study bacteria, while one investigates a virus. Two studies utilize polymerase chain reaction (PCR), standard phenotypic microbiological methods, whole-genome sequencing (WGS), and a combination of both standard methods and PCR. One study analyzes laboratory data from foodborne outbreaks, and utilizes Matrix-Assisted Laser Desorption–Ionization Time-of-Flight Mass Spectrometry (MALDI-TOF). The broad spectrum of methods used to detect and characterize the pathogens in the studies reflects the availability of expertise, facilities, and funds; it is significant that the objectives of the investigations are achieved. However, the fact that six of the ten studies employ molecular methods (PCR and WGS) reflects the current trends in the application of tools, which the audience may find interesting. It cannot be over-emphasized that it is vital to make new findings and knowledge available to scientists and the general public. To achieve this objective, I seize this opportunity to express my gratitude to the MDPI book staff responsible for the production of the e-book of our Special Issue, entitled "Foodborne Pathogens: Detection Methods, Food Safety, and Public Health". I also appreciate the invaluable contributions of the Editors of Microorganisms and the authors of the ten papers published in this Special Issue.

Abiodun Adewale Adesiyun
Editor

microorganisms

MDPI

Communication

Study on the Effect of Relaying on Norovirus Reduction from *Crassostrea gigas* Oysters

Roberta Battistini [1,*], Chiara Masotti [1], Cristiana Maurella [1], Erica Costa [2], Mino Orlandi [2], Mirvana Feletti [3], Carlo Ercolini [1] and Laura Serracca [1]

1 Istituto Zooprofilattico Sperimentale del Piemonte, Liguria e Valle d'Aosta, 10154 Turin, Italy
2 Liguria Local Health Unit-ASL 5, Complex Unit of Hygiene of Foods and Animal Origin, 19122 La Spezia, Italy
3 Dipartimento Agricoltura, Turismo, Formazione e Lavoro Regione Liguria—Settore Politiche Agricole e Della Pesca, 16121 Genoa, Italy
* Correspondence: roberta.battistini@izsto.it

Abstract: Norovirus (NoV) is the most important cause of seafood-borne gastroenteritis worldwide, mainly associated with the consumption of raw oysters. NoV is often present in oysters that comply with existing control standards for shellfish. Therefore, the improvement of post-harvest treatments and practices can represent one of the main strategies to reduce the incidence of viral diseases related to shellfish. This study aimed to investigate long-term relays for the reduction of NoV levels in live oysters, during the high-risk cold months, by transferring the oysters from a more contaminated site to two sites with lower NoV levels. The efficacy of relaying was evaluated by analyzing oyster samples collected at days 0 (T0) and 30 (T30) for NoV levels using a real-time quantitative PCR (RT-qPCR). The NoV level at the relay sites was consistently lower than at the harvest site. The relay process for 30 days in seawater with a lower NoV level resulted in a decrease in the NoV load compared to day 0 with significant reductions depending on the site and genogroup of NoV considered. These results suggest that long-term relaying of oysters to reduce NoV levels is promising and could help growers to improve oyster safety; however, further investigations are needed.

Keywords: *Crassostrea gigas*; relay; norovirus; food safety; real-time qPCR

Citation: Battistini, R.; Masotti, C.; Maurella, C.; Costa, E.; Orlandi, M.; Feletti, M.; Ercolini, C.; Serracca, L. Study on the Effect of Relaying on Norovirus Reduction from *Crassostrea gigas* Oysters. *Microorganisms* **2022**, *10*, 2389. https://doi.org/10.3390/microorganisms10122389

Academic Editor: Abiodun Adewale Adesiyun

Received: 28 September 2022
Accepted: 30 November 2022
Published: 1 December 2022

1. Introduction

Oysters are filter-feeding organisms capable of concentrating pathogenic microorganisms from the environment in which they live. The widespread habit of consuming these raw shellfish exposes the consumer to the risk of gastroenteritis. Among the main pathogens responsible for foodborne gastroenteritis are Noroviruses (NoVs) which caused 130 outbreaks in Europe in 2020 [1]. The latest European Food Safety Authority (EFSA) study in 2021, which measured NoV contamination in raw oysters, reported that one-third of mussels in Europe are contaminated. NoV was found in 34.5% of mussels collected at production sites and in 10.8% of those for sale. The survey showed higher NoV contamination in the period from November to April, as well as lower contamination for class A areas compared to class B and class C areas. In the class A areas, the laws do not provide for particular treatments before marketing, which is necessary for oysters from the class B areas [2]. NoV infection is prevalent from November to April and is sometimes called "winter vomiting disease" or "stomach flu". The virus is highly infectious, and ten viral particles are enough to give rise to an infection [3]. NoV is transmitted mainly through the fecal–oral route, by consumption of contaminated food or water, or directly from person to person, and also by contact with contaminated surfaces. NoV can be excreted in high levels (up to 10^9 viruses/g feces) in the feces of infected individuals [4]. Therefore, during the winter period, high concentrations of NoV can be found in wastewater [5]. Wastewater discharge into aquatic environments is practiced worldwide, representing an important issue in coastal seawaters during the winter and leading to the contamination of bivalve mollusk production areas. In particular, oysters contaminated with NoV pose an important risk to human health since they are usually consumed raw [6].

The conventional approach to the purification of oysters envisaged by current legislation is depuration, a process used to remove microorganisms and other contaminants from bivalve mollusks by placing them in tanks of clean seawater, often recycled and disinfected using ultraviolet light, ozone, or other means [3]. Depuration is used worldwide, and the depuration periods are varied from a few hours to several days, depending on the country [7]. The effectiveness of the depuration process is evaluated by the ability to reduce the bacterial count, often using fecal coliform bacteria and, in particular, the "target" microorganism *E. coli*. Depuration of shellfish is not effective for reducing enteric viruses, and episodes of viral gastroenteritis associated with the consumption of bivalve mollusks evaluated with the *E. coli* parameter occur annually around the world [8]. It is therefore important to identify different post-harvest intervention strategies to reduce these pathogens in oysters in order to increase their safety for consumers [9].

Relaying is an alternative treatment to depuration; this involves a longer-term purification process (often requiring ten days or longer) [10]. In relaying, bivalve mollusks are collected from a contaminated area and transferred to pollution-free marine environments, where they are maintained for a shorter or longer period to allow them to purge contaminants derived from wastewater under natural environmental conditions.

Limited data on the effectiveness of prolonged relaying for NoV reduction are currently available [11–14], and further studies about this topic are needed, preferably using the standardized European Committee for Standardization (CEN) method [9]. So far, two studies have suggested that a relaying period in seawater of around four weeks may be sufficient to reduce NoV levels below the limit of quantification (LOQ) in oysters [11,13]. Unfortunately, very few marine areas are completely NoV-free, and also the waters classified as class A are contaminated by NoV. Therefore, our study evaluated the reduction of NoV concentration in the winter months using the quantitative real-time RT-PCR method to further investigate the long-time relaying period by relaying oysters for 30 days in seawater sites with less NoV contamination.

2. Materials and Methods

2.1. Oyster Sampling Sites

Oysters (*Crassostrea gigas*) used in this study were harvested from a class B shellfish farming area in northwestern Italy (site 1) and moved to two different sites of category A (site 2 and site 3) (Figure 1). Site 1 is located inshore in a very anthropized area characterized by a commercial and tourist port and shipbuilding activity. Sites 2 and 3 are offshore at approximately 1.6 km and 1.8 km from the coast, respectively (Figure 1). Sixty oysters were harvested from a single sampling point in the main production area (site 1) every month, for a total of four samplings, during the coldest months of the year (November–February) when NoV concentration is supposed to be the highest. Twenty oysters were immediately analyzed (ten individuals tested twice) for NoV genogroup I (GI) and genogroup II (GII) (T0) from each sampling. The other oysters were moved to site 2 and site 3 with less NoV contamination and left for relaying in these seawaters for one month. After 30 days, the 20 oysters from site 2 and the 20 oysters from site 3 were collected, transferred to the laboratory in refrigerated condition, and immediately analyzed for NoV GI and GII (10 individuals tested twice for each site) (T30).

Additionally, samples of 20 oysters were harvested from sites 2 and 3 on the same days as the oyster sampling at site 1 (T0). Further samples of 20 oysters were sampled from sites 2 and 3 on the same days as the oyster sampling after 30 days of relaying (T30). These additional samples were analyzed for NoV GI and NoV GII to monitor the concentration of NoV in native oysters. Seawater parameters, such as temperature and salinity, were recorded at each sampling in site 2 and 3.

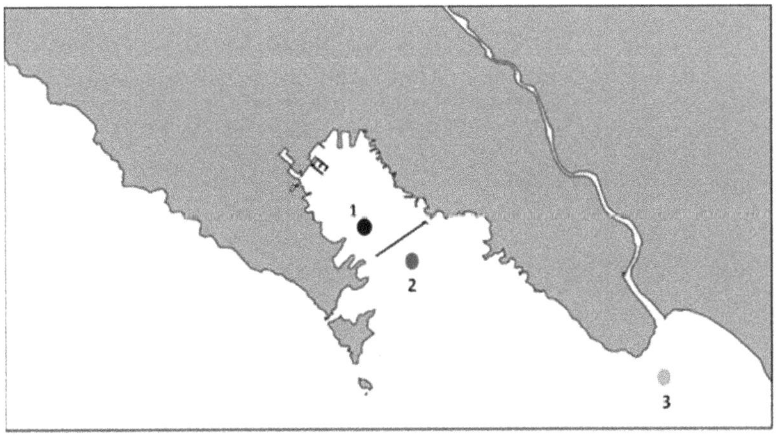

Figure 1. Schematic map of sampling sites: (1) Main production area, (2) relaying site 2, (3) relaying site 3. The approximate distance from (1) to (2) is 0.94 Km, and (1) to (3) is 11.5 Km.

2.2. Virus Recovery from Oysters

Oyster samples (each composed of ten oysters) were tested according to the ISO 15216-1:2017 method [15]. The hepatopancreas (on average 5 g per oyster) was removed by dissection from each oyster, pooled, and homogenized with TissueLyser (Qiagen, Hilden, Germany). A total of 10 μL of Mengovirus (process control virus) and 2 mL of proteinase K (0.1 mg/mL) were added to 2 grams of homogenates. The homogenates were incubated for 60 min at 37 °C with shaking at 320 rpm and, after that, maintained at 60 °C for 15 min in the water bath and centrifuged for 5 min at 3000× *g*. Finally, supernatants were recovered for RNA extraction, and their volumes were recorded.

2.3. RNA Extraction

Viral RNA was extracted from 500 μL of the supernatants using the EGENE-UP® platform and the NucliSens magnetic extraction reagents (bioMérieux, Marcy l'Etoile, France) according to the manufacturer's instructions. RNA was eluted into 100 μL of elution buffer and was immediately used for the NoV quantification by real-time RT-PCR or stored at −80 °C until real-time RT-PCR analysis.

2.4. Quantification of Norovirus by Real-Time RT-PCR

The detection and quantification of NoV GI and NoV GII genomes were performed by real-time quantitative reverse transcription PCR (RT-qPCR) according to ISO 15216-1:2017, and the reactions were carried out using the Biorad CFX96TM Real-Time PCR thermocycler (Biorad, Hercules, CA, USA). The number of RNA copies per μL of each sample was calculated by matching the sample Cq value to the standard curves (one for each target) created with the tenfold serial dilution of a dsDNA standard for NoV GI and NoV GII. Therefore, the final concentrations were expressed as genomic copies per gram (g.c./g) and were calculated based on the volume of the analyzed extract. According to ISO 15216-1:2017, the Cq value of Mengovirus was obtained in spiked samples and was compared with extracted samples by viral stock to evaluate extraction efficiency. Furthermore, to evaluate the inhibition of RT-qPCR, the Cq value was obtained in samples spiked with 1 μL of external control RNA for both GI and GII and was compared with that obtained in samples without the addition of external control. Results with extraction efficiency > 1% and RT-qPCR inhibition ≤ 75% were considered valid. The LOQ was established by the European Union Reference Laboratory (EURL). The LOQ for NoV GI was calculated as 140 g.c./g and 130 g.c./g for NoV GII.

2.5. Statistical Analysis

Data were verified for Normality using the Shapiro–Wilk test. The Wilcoxon rank-sum test was used to compare the samples between the two oyster farming areas at time T0. The Wilcoxon matched-pairs signed-rank test was used to compare the equality of matched pairs of observations, i.e., time T0 versus T30.

3. Results

During the study period from November to February, NoV GI and NoV GII were always detected at harvest site 1 in a concentration higher than sites 2 and 3, except for NoV GI in December month (second sampling), when it was present at higher concentration at site 2. NoV GII was found at all sites at higher concentrations than NoV GI. The maximum NoV GII load was 3.1×10^5 viral g.c./g at site 1, 5.5×10^4 viral g.c./g at site 2, and 2.7×10^4 viral g.c./g at site 3. The maximum load of NoV GI was 5.1×10^2 viral g.c./g at site 1, 1.9×10^3 viral g.c./g at site 2, and 6.2×10^3 viral g.c./g at site 3. Moreover, NoV GII was always detected in all sampling, while NoV GI was absent or below the LOQ in some samples at sites 2 and 3. Among the sites used for relaying, site 3 was the one with a lower concentration of both NoV GI and NoV GII compared to site 2 (Figure 2a,b).

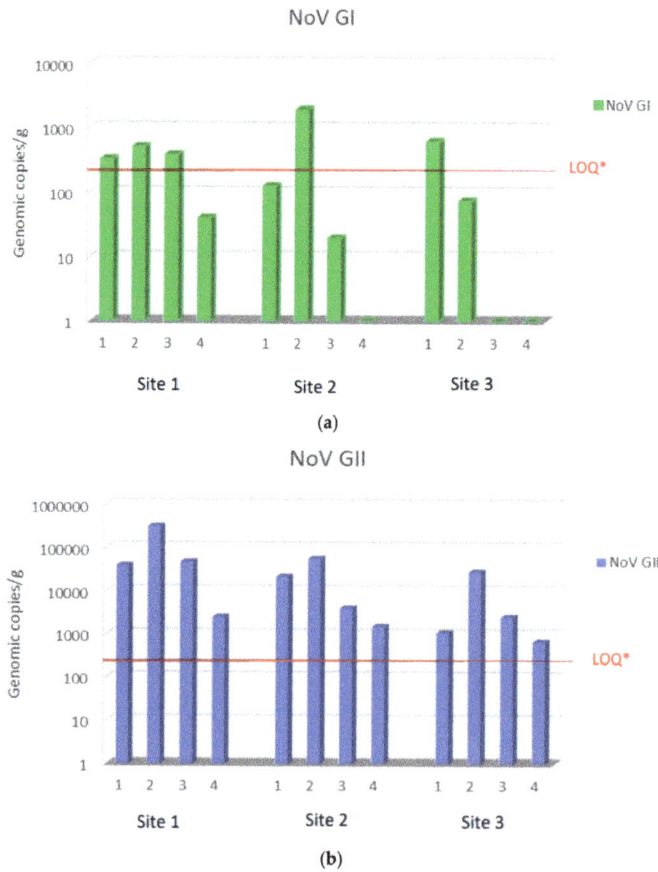

Figure 2. NoV RNA levels detected in oysters at Time 0 of the relay (T0) at site 1, site 2, and site 3: (**a**) NoV GI; (**b**) NoVGII. Values are expressed in genome copies/gram. * LOQ: 140 g.c./g for NoV GI and 130 g.c./g for NoV GII.

NoV GI levels decreased in all samples below the LOQ of the assay (140 genomic copies per g) at site 2 after 30 days from the transfer of oysters from site 1. Moreover, NoV GI was not detected in two samples after relaying period (Figure 3a). In the same site, NoV GII levels decreased in all oyster samples except in the first sampling, while NoV GII levels decreased below the LOQ (130 genomic copies per g) in the last sampling (Figure 3b).

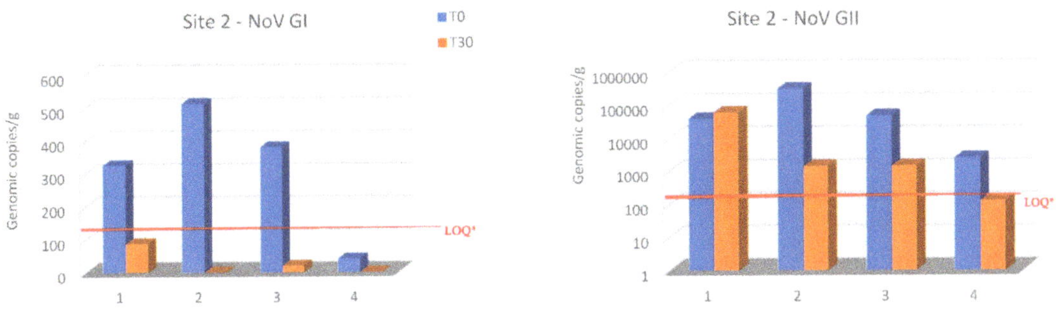

(a) (b)

Figure 3. Comparison of NoV RNA levels in oysters at Time 0 of the relay (T0) and after 30 days of the relay (T30) at site 2: (**a**) NoV GI; (**b**) NoV GII. Values are expressed in genome copies/gram. * LOQ: 140 g.c./g for NoV GI and 130 g.c./g for NoV GII.

Both NoV GI and NoV GII levels decreased in all oyster samples of site 1 when transferred to site 3 after 30 days of relaying (Figure 4a,b), and NoV GI was no longer detected after this period (Figure 4a). NoV GII was below the LOQ in samples collected in the second and fourth sampling (December and February) (Figure 4b). Unfortunately, it was not possible to evaluate the samples placed at site 3 in November because the samples were not found after 30 days, probably due to theft by unknown persons.

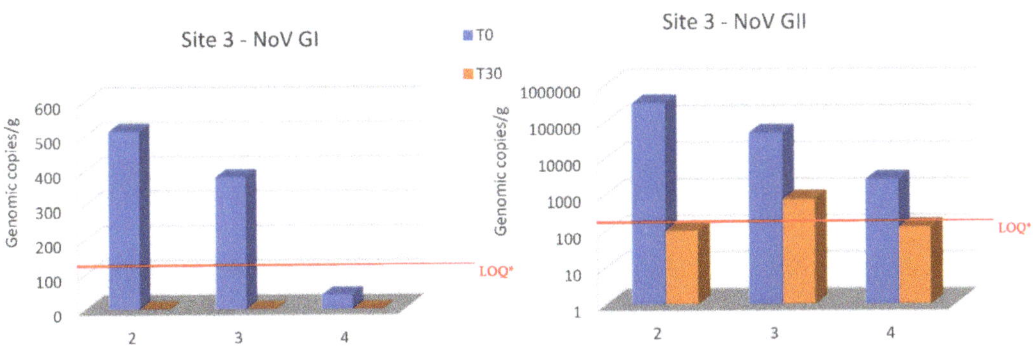

(a) (b)

Figure 4. Comparison of NoV RNA levels in oysters at Time 0 of the relay (T0) and after 30 days of the relay (T30) at site 3: (**a**) NoV GI; (**b**) NoV GII. Values are expressed in genome copies/gram. * LOQ: 140 g.c./g for NoV GI and 130 g.c./g for NoV GII.

Relays for 30 days in seawater with a lower NoV level resulted in a decrease in the NoV load in oysters compared to T0 at both sites 2 and 3. In particular, logarithmic reductions between 2.34 and −0.2 and between 2.7 and 0.56 were obtained, respectively, for NoV GII and NoV GI at site 2 (Figure 3). Although, log reductions ranged from 3.48 to 1.3 for NoV GII and from 2.7 to 1.6 for NoV GI at site 3 (Figure 4). After 30 days of relaying, the load of NoV in oysters usually tends to conform to the area where they are introduced (Figure 5).

Based on statistical analysis, no difference between the two groups was found at time T0 but a statistically significant difference was found by comparing the withdrawal data both for NoV GI (p = 0.04) and for Nov GII (p = 0.009). A statistically significant difference was found by comparing the paired data between time T0 and time T30 both for NoV GI (p = 0.001) and for Nov GII (p = 0.01). The comparison between the two consecutive summations also gave a significant result (p = 0.01). The statistical analysis of stratified sampling at relaying area gave the following results: at site 2, only the reduction of NoV GI was significant (p = 0.031), while at site 3, both NoV GII and the sum of NoV GI and NoV GII reduction were significant (p = 0.031).

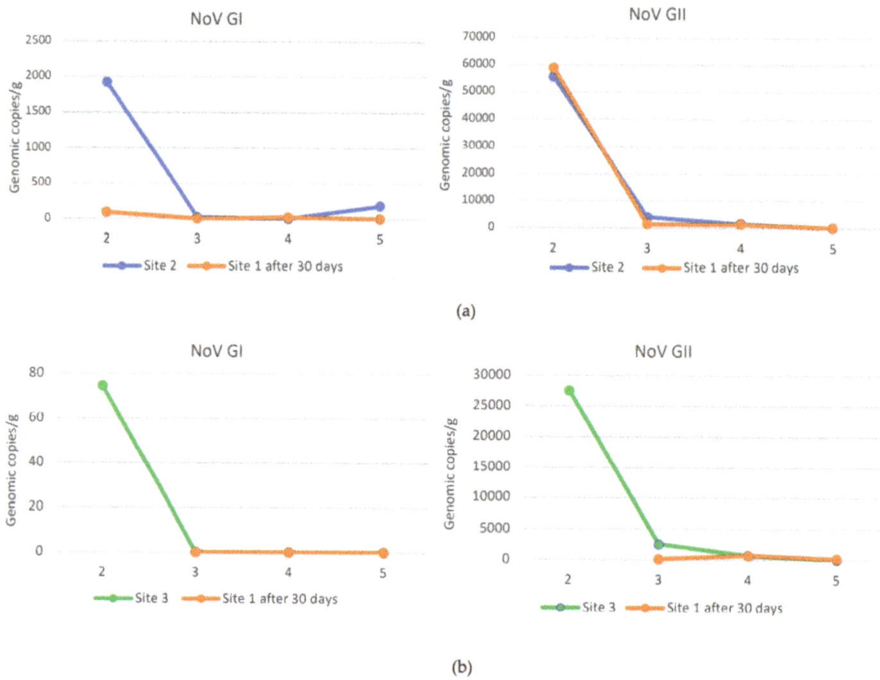

Figure 5. NoV GI and NoV GII values in oysters at site 1 after 30 days of relaying at site 2 (**a**) or site 3 (**b**). NoV GI and NoV GII values in oysters naturally presented at site 2 (**a**) or site 3 (**b**) at T30.

Physiochemical parameters of seawater during oysters' relay periods are summarized in Table 1. Temperature and salinity showed no significant variations between the two sites in the same months.

Table 1. Physiochemical parameters of seawater measured during oysters' relaying period at sites 2 and 3.

	Site 2		**Site 3**	
	T (°C) Media ± dev. st.	**Salinity (PSU)**	**T (°C) Media ± dev. st.**	**Salinity (PSU)**
November	17.0 ±1.4	37.5 ± 0.5	17.2 ± 1.2	36.7 ± 0.7
December	13.6 ±1.1	37.3 ± 0.6	12.6 ± 1.0	36.2 ± 0.5
January	13.8 ±1.0	37.8 ± 0.4	12.4 ± 0.6	36.7 ± 0.4
February	12.8 ± 0.5	37.8 ± 0.6	12.9 ± 1.1	37.5 ± 0.5
March	13.4 ± 0.85	38.0 ± 0.5	13.5± 0.9	37.6 ± 0.6

4. Discussion

In this study, we report the results of prolonged relaying to decrease NoV concentration in oysters as a possible management strategy to reduce consumer exposure to NoV in the winter season. To study NoV reduction, we used environmentally contaminated oysters from approved production areas classified as Class B areas and transferred them to two different sites with less contamination classified as Class A during the high-risk winter period. NoV monitoring at the main production sites (site 1) and the relaying sites (sites 2 and 3) confirmed that oysters at the main production site contained, on average, higher NoV concentrations.

In our study, the presence of high levels of NoV at site 1, compared to sites 2 and 3, is probably related to the characteristics of the area where the oysters are raised. The oyster farm located at site 1 is inside a dam, which protects the commercial port from the open sea by reducing sea currents and favoring the stagnation of contaminants in this area. Whereas in oyster farms located at sites 2 and 3, the depth of the seabed and the distance from the coast favor the dilution and quantity of contaminants present in marine waters and reduce the risk of contamination of oysters.

The data presented here indicate that oysters from the main collection site significantly reduced their NoV concentration by relaying for 30 days in seawater sites with a lower NoV concentration. In only one case, the NoV GII concentration increased after 30 days at site 2 (Figure 3b). The increased concentration can be explained by analyzing the NoV GII level found in the second sampling in this area. In fact, when sample 1 was placed at site 2 (T0) in November, the NoV concentration was lower than that present at the re-immission site. During the following 30 days, the NoV load at site 2 increased, as seen in Figure 2b. Therefore, in this case, the sample moved from site 1 to site 2 could not reduce the NoV load present in it during the relaying period. Another unexpected result is related to the first sample from site 1 introduced into site 2, containing after 30 days a lower load of NoV GI than the oysters always present at site 2 (Figure 5a). These data are related to the second sampling carried out in December, when evidently there was a peak of both NoV GI and NoV GII in this area. This case highlights a different behavior of the two NoV genogroups in oysters, which bind to different ligands within the tissues of the oysters and, therefore, are concentrated and/or eliminated differently, as already highlighted in other studies [16,17]. Except for this case, the NoV values after the relaying period tend to conform to those values recorded in oysters already present at sites 2 and 3 (Figure 5). Indeed, the best NoV reduction value was obtained in the oyster samples after 30 days at site 3, which maintained a lower NoV load during the winter period. In detail, in 67% of the oyster samples, NoV was <LOQ at site 3, while at site 2, only 25% of the samples were below the LOQ. NoV GI was reduced at both sites in all samples below the LOQ, while the NoV GII was reduced more at site 3, with 50% of the samples < LOQ after 30 days of relays in adherence to the situation present in the relaying areas. No threshold infectivity limit is currently established for NoV as detected by polymerase chain reaction (PCR). However, it is shown that a low probability of outbreaks is associated with oysters containing NoV in concentration levels below 152 g.c./g., which in our study are present in samples below the LOQ. However, it is not possible to conclusively rule out the possibility of oysters containing levels of <152 c.g./g. causing illness, and it seems likely that these levels present a comparatively lower risk. In addition, there is some indication that at higher levels (>500 c.g./g.), the risk becomes greater [18].

So far, few studies have reported a significant reduction in NoV load from oysters relayed in seawater over an extended period. In one case, oysters from the harvesting area responsible for NoV outbreaks were moved to a seawater site free from sewage contamination and maintained there for 17 days [13]. In another study, the oysters reduced NoV concentration by <500 genomic copies/g in all samples relayed during the winter season in an alternative site with less NoV contamination. In contrast, 31% of oyster samples kept at the native harvest site contained NoV > 500 genomic copies/g [14].

The ideal procedure for obtaining oysters safely would be harvesting them in areas that are not subjected to any type of contamination. However, very few such areas really exist, and also the waters classified as class A are contaminated by NoV. Moreover, access to clean relaying waters may be a challenge in the future, with an increasing global population. To overcome these problems, long-term relaying could be a mitigation strategy for the enteric virus reduction; if possible, move the oysters from areas with high levels of NoV to areas with a lower viral concentration in order to reduce the risk to consumers as much as possible. Indeed, in some studies, it is observed that the likelihood of becoming infected with NoV increases with NoV dose [4,19–21]. Furthermore, a correlation was found between the amount of self-reported disease and the number of copies of the NoV genome in oysters [18,22]; higher concentrations of NoV RNA correspond to a higher rate of reported diseases, suggesting a link between virus RNA levels and health risks. Currently, the most widespread method to reduce contamination is depuration, with good results in the elimination of fecal bacteria but scarcely or not all effective in eliminating pathogenic viruses. The other post-harvest treatments, such as frozen storage, thermal inactivation, and high-pressure processing, also require either a significant amount of initial investment or operation costs and often change the organoleptic characteristics of shellfish, making them unacceptable to consumers [23]. The application of long-term relaying instead could provide a practical, less expensive, and natural alternative to other methods by enabling the reduction of NoV concentrations to levels that reduce, if not eliminate, the risk to consumers. Long-term relaying is especially important for producers that can move the oysters to less contaminated sites close to other more contaminated production areas. Considering that the oyster takes 18 to 24 months to become an adult or reach market size (about 3 inches), it would be a matter of moving oysters to the relaying sites one month before being sold, in the winter season only, so to minimize the extra costs related to the relocation process.

In our study, the obtained dataset provides statistically significant differences between the different sites studied and between the different genogroups and gives an initial indication of the trends and effects of this treatment. However, further studies during winter seasons may be useful to integrate the data presented here. The limitation of our study was the lack of Norovirus-free marine areas to move the oysters to for relaying since this strategy would be the best for studying the oyster relaying. Unfortunately, in our case, no NoV-free marine areas are close to the main production site where the oysters could be transferred at affordable costs before sale. As discussed above, it is increasingly difficult to find NoV-free marine waters, and, as in our case, waters classified in class A are also usually contaminated by NoV, albeit in a lower concentration than the sites classified in class B. Therefore, in our study, we chose to move the oysters from a more NoV-contaminated site to less NoV-contaminated sites in order to study the long-term relay for the reduction of NoV concentrations in oysters.

5. Conclusions

In conclusion, the results of this study suggest that relaying naturally contaminated oysters to sites close to the main production area with lower NoV concentration before their final harvest could provide a practical and low-cost mitigation strategy to reduce NoV in oysters. Relaying could lead to a decrease in the risk of NoV disease from oyster consumption during the colder season when NoV is present in high concentrations. This study may help growers to improve both the marketability and safety of oysters, as well as provide additional information relevant to risk management decisions for regulatory agencies in light of the potential introduction of a statutory limit.

Author Contributions: Conceptualization, C.E., M.O. and L.S.; methodology, R.B., C.M. (Chiara Masotti), E.C. and L.S.; formal analysis, R.B. and C.M. (Chiara Masotti); investigation, R.B., C.M. (Chiara Masotti), E.C. and L.S.; resources, L.S., C.E. and E.C.; data curation, R.B., C.M. (Chiara Masotti) and C.M. (Cristiana Maurella); writing— original draft preparation, R.B. and C.M. (Chiara Masotti); writing—review and editing, R.B. and C.M. (Chiara Masotti); supervision, C.E., M.O., M.F. and L.S.; project administration, L.S. and M.F.; funding acquisition, L.S. and M.F. All authors have read and agreed to the published version of the manuscript.

Funding: This research was funded by the Italian Ministry of Health, grant number IZS PLV 13/19 RC and by Liguria Region FEAMP 2104/2020 mis. 2.51 grant number 19E05 the APC was funded by IZS PLV 13/19 RC.

Data Availability Statement: Not applicable.

Acknowledgments: The authors would like to thank Mitilicoltori Associati S.C.a r.l.-Lerici (SP) Italy and ARPAL for their collaboration in oyster sampling and for the physiochemical parameters of the seawater supplied. We would also like to thank Pite Ledi for the English language review.

Conflicts of Interest: The authors declare no conflict of interest.

References

1. European Food Safety Authority; European Centre for Disease Prevention and Control. The European Union One Health 2020 Zoonoses Report. *EFSA J.* **2021**, *19*, 6971.
2. European Food Safety Authority. Analysis of the European baseline survey of norovirus in oysters. *EFSA J.* **2019**, *17*, e05762.
3. Yezli, S.; Otter, J.A. Minimum infective dose of the major human respiratory and enteric viruses transmitted through food and the environment. *Food Environ. Virol.* **2011**, *3*, 1–30. [CrossRef] [PubMed]
4. Teunis, P.F.; Sukhrie, F.H.; Vennema, H.; Bogerman, J.; Beersma, M.F.; Koopmans, M.P. Shedding of norovirus in symptomatic and asymptomatic infections. *Epidemiol. Infect.* **2015**, *143*, 1710–1717. [CrossRef] [PubMed]
5. Corpuz, M.V.A.; Buonerba, A.; Vigliotta, G.; Zarra, T.; Ballesteros, F., Jr.; Campiglia, P.; Belgiorno, V.; Korshin, G.; Naddeo, V. Viruses in wastewater: Occurrence, abundance and detection methods. *Sci. Total Environ.* **2020**, *745*, 140910. [CrossRef] [PubMed]
6. Flannery, J.; Keaveney, S.; Rajko-Nenow, P.; O'Flaherty, V.; Doré, W. Concentration of norovirus during wastewater treatment and its impact on oyster contamination. *Appl. Environ. Microbiol.* **2012**, *78*, 3400–3406. [CrossRef] [PubMed]
7. Lee, R.; Lovatelli, A.; Ababouch, L. Bivalve Depuration: Fundamental and Practical Aspects. 2008, pp. 1–139. Available online: http://www.fao.org/3/i0201e/i0201e.pdf (accessed on 5 September 2022).
8. McLeod, C.; Polo, D.; Le Saux, J.C.; Le Guyader, F.S. Depuration and Relaying: A Review on Potential Removal of Norovirus from Oysters. *Compr. Rev. Food Sci. Food Saf.* **2017**, *16*, 692–706. [CrossRef] [PubMed]
9. European Food Safety Authority. Scientific opinion on Norovirus (NoV) in oysters: Methods, limits and control options. *EFSA J.* **2012**, *10*, 2500. [CrossRef]
10. Lees, D.; Younger, A.; Dore, B. Depuration and relaying. In *Safe Management of Shellfish and Harvest Waters*; Rees, G., Pond, K., Kay, D., Bartram, J., Santo Domingo, J., Eds.; IWA Publishing: London, UK, 2010; Volume 9, pp. 145–181.
11. Le Guyader, F.S.; Le Saux, J.C.; Ambert-Balay, K.; Krol, J.; Serais, O.; Parnaudeau, S.; Giraudon, H.; Delmas, G.; Pommepuy, M.; Pothier, P.; et al. Aichi virus, norovirus, astrovirus, enterovirus, and rotavirus involved in clinical cases from a French oyster-related gastroenteritis outbreak. *J. Clin. Microbiol.* **2008**, *46*, 4011–4017. [CrossRef] [PubMed]
12. Greening, G.; Hewitt, J.; Hay, B.; Grant, C. Persistence of Norwalk-like viruses over time in Pacific oysters grown in the natural environment. In Proceedings of the 4th International Conference on Molluscan Shellfish Safety, Santiago de Compostela, Spain, 4–8 June 2002.
13. Doré, W.J.; Keaveney, S.; Flannery, J.; Rajko-Nenow, P. Management of health risk associated with oysters harvested from a norovirus contaminated area, Ireland, February–March 2010. *Euro Surveill.* **2010**, *15*, 1–5. [CrossRef]
14. Rupnik, A.; Keaveney, S.; Devilly, L.; Butler, F.; Doré, W. The impact of winter relocation and depuration on norovirus concentrations in pacific oysters harvested from a commercial production site. *Food Environ. Virol.* **2018**, *10*, 288–296. [CrossRef] [PubMed]
15. *ISO 15216-1:2017*; Microbiology of Food and Animal Feed—Horizontal Method for Determination of Hepatitis A Virus and Norovirus in Food Using Real time RT-PCR—Part1: Method for Quantification. International Organization for Standardization: Geneva, Switzerland, 2017.
16. Le Guyader, F.; Loisy, F.; Atmar, R.L.; Hutson, A.M.; Estes, M.K.; Ruvoën-Clouet, N.; Pommepuy, M.; Le Pendu, J. Norwalk virus-specific binding to oyster digestive tissues. *Emerg. Infect. Dis.* **2006**, *12*, 931–936. [CrossRef] [PubMed]
17. Le Guyader, F.S.; Atmar, R.L.; Le Pendu, J. Transmission of viruses through shellfish: When specific ligands come into play. *Curr. Opin. Virol.* **2012**, *2*, 103–110. [CrossRef] [PubMed]
18. Lowther, J.A.; Gustar, N.E.; Hartnell, R.E.; Lees, D.N. Comparison of norovirus RNA levels in outbreak-related oysters with background environmental levels. *J. Food Prot.* **2012**, *75*, 389–393. [CrossRef] [PubMed]
19. TerWaarbeek, H.L.; Dukers-Muijrers, N.H.; Vennema, H.; Hoebe, C.J. Waterborne gastroenteritis outbreak at a scouting camp caused by two norovirus genogroups: GI and GII. *J. Clin Virol.* **2010**, *47*, 268–272. [CrossRef] [PubMed]
20. Teunis, P.F.M.; Le Guyader, F.S.; Liu, P.; Ollivier, J.; Moe, C.L. Noroviruses are highly infectious but there is strong variation in host susceptibility and virus pathogenicity. *Epidemics* **2020**, *32*, 100401. [CrossRef] [PubMed]
21. Visser, H.; Verhoef, L.; Schop, W.; Götz, H.M. Outbreak investigation in two groups of coach passengers with gastroenteritis returning from Germany to the Netherlands in February 2009. *Euro Surveill.* **2010**, *15*, 19615. [CrossRef] [PubMed]

22. Lowther, J.A.; Avant, J.M.; Gizynski, K.; Rangdale, R.E.; Lees, D.N. Comparison between quantitative real-time reverse transcription PCR results for norovirus in oysters and self-reported gastroenteric illness in restaurant customers. *J. Food Prot.* **2010**, *73*, 305–311. [CrossRef] [PubMed]
23. Gyawali, P.; Fletcher, G.C.; McCoubrey, D.J.; Hewitt, J. Norovirus in shellfish: An overview of post-harvest treatments and their challenges. *Food Control* **2019**, *99*, 171–179. [CrossRef]

microorganisms

Article

Evaluation of Various Lactic Acid Bacteria and Generic *E. coli* as Potential Nonpathogenic Surrogates for In-Plant Validation of Biltong Dried Beef Processing

Caitlin E. Karolenko [1,2], Jade Wilkinson [1,2] and Peter M. Muriana [1,2,*]

[1] Robert M. Kerr Food and Agricultural Product Center, Oklahoma State University, Stillwater, OK 74078, USA
[2] Department of Animal and Food Sciences, Oklahoma State University, Stillwater, OK 74078, USA
[*] Correspondence: peter.muriana@okstate.edu; Tel.: +1-405-744-5563

Abstract: Validation studies conducted within a food processing facility using surrogate organisms could better represent the manufacturing process than controlled laboratory studies with pathogenic bacteria on precision equipment in a BSL-2 lab. The objectives of this project were to examine potential surrogate bacteria during biltong processing, conduct biltong surrogate validation lethality studies, and measure critical factors and intrinsic parameters during processing. Beef pieces (1.9 cm × 5.1 cm × 7.6 cm) were inoculated with four-strain mixtures of *Carnobacterium divergens/C. gallinarum*, *Pediococcus acidilactici/P. pentosaceous*, and Biotype 1 *E. coli* ATCC BAA (-1427, -1428, -1429, and -1430), as well as a two-strain mixture of *Latilactobacillus sakei* and other commercially available individual bacterial cultures (*P. acidilactici* Saga200/Kerry Foods; *Enterococcus faecium* 201224-016/Vivolac Cultures). Inoculated beef was vacuum-tumbled in marinade and dried in a humidity-controlled oven for 8–10 days (24.9 °C; 55% relative humidity). Microbial enumeration of surviving surrogate bacteria and evaluation of intrinsic factors (water activity, pH, and salt concentration) were performed post inoculation, post marination, and after 2, 4, 6, 8, and 10 days of drying. Trials were performed in duplicate replication with triplicate samples per sampling time and analyzed by one-way RM-ANOVA. Trials conducted with *E. faecium*, *Pediococcus* spp., and *L. sakei* never demonstrated more than 2 log reduction during the biltong process. However, *Carnobacterium* achieved a >5 log (5.85 log) reduction over a drying period of 8 days and aligned with the reductions observed in previous trials with pathogenic bacteria (*Salmonella, E. coli* O157:H7, *L. monocytogenes*, and *S. aureus*) in biltong validation studies. Studies comparing resuspended freeze-dried or frozen cells vs. freshly grown cells for beef inoculation showed no significant differences during biltong processing. *Carnobacterium* spp. would be an effective nonpathogenic in-plant surrogate to monitor microbial safety that mimics the response of pathogenic bacteria to validate biltong processing within a manufacturer's own facility.

Keywords: biltong; surrogate; lactic acid bacteria; dried beef; validation; *Carnobacterium*

Citation: Karolenko, C.E.; Wilkinson, J.; Muriana, P.M. Evaluation of Various Lactic Acid Bacteria and Generic *E. coli* as Potential Nonpathogenic Surrogates for In-Plant Validation of Biltong Dried Beef Processing. *Microorganisms* **2022**, *10*, 1648. https://doi.org/10.3390/microorganisms10081648

Academic Editors: Salam A. Ibrahim and Abiodun Adewale Adesiyun

Received: 20 June 2022
Accepted: 12 August 2022
Published: 15 August 2022

1. Introduction

Biltong is a South African style dried beef product that is growing in popularity in the United States. This dried meat product is traditionally made using lean strips of beef that are marinated in a mixture of traditional spices (coriander and pepper), salt, and vinegar and then dried at low or ambient temperature and humidity. Dried beef processing guidelines, as issued by the United States Department of Agriculture Food Safety and Inspection Service (USDA-FSIS), require dried beef products to be heated to an internal temperature of 160 °F (71.1 °C) in a sealed oven or steam injector with a relative humidity greater than 90% during the cooking/heating process [1]. Since biltong does not have a heat lethality step during processing and deviates from these guidelines, biltong manufacturers must conduct a validation or challenge study to evaluate the ability of their process to sufficiently inactive bacterial pathogens such as *Salmonella* spp. which have been historically linked to outbreaks and recalls of dried meat and poultry products [2]. USDA-FSIS does give processors two different options to safely produce these alternative dried meat products. The first option requires *Salmonella* testing of every lot of edible ingredients used during processing and an overall process reduction of a 'pathogen of concern' of at least 2 log. Alternatively, processors can forego ingredient testing if they can demonstrate that their process can achieve $\geq$ 5 log reduction of *Salmonella* by the end of processing [3].

USDA-FSIS regulatory guidance for manufacture and sale of biltong requires processors to demonstrate product safety by process validation against a 'pathogen of concern'. In recent BSL-2 in-lab studies, this was performed with *Salmonella* serovars [4], *E. coli* O157:H7, *L. monocytogenes*, and *S. aureus* [5]. These experiments, while successful in achieving a >5 log reduction of foodborne pathogens where the data are currently used by processors in support of their in-plant food safety (HACCP) processes, are often conducted in highly controlled BSL-2 laboratory environments with research-grade equipment. The food processing environment is extremely variable between small and large processors, and both likely have greater variability of process parameters than that found in BSL-2 lab equipment. USDA-FSIS has recognized this difference and has allowed consideration of 'in-plant' validation studies using surrogate organisms if the surrogate can mimic a pathogen's response to a process [6–8]. The intention is that in-plant data would more likely reflect the actual process variability and conditions than scientific equipment from a BSL-2 lab. Conducting a validation study within a processor's own facility would allow for a more accurate representation of the impact of a commercial process on pathogenic bacteria. Due to food safety concerns, it is unsafe to introduce pathogenic bacteria into a manufacturing facility to test whether the process achieves sufficient microbial reduction. Therefore, nonpathogenic surrogate bacteria would be better suited to mimic the response of pathogens to actual processing conditions [8]. This presents the following question: what surrogate organism should be used for the biltong process?

A surrogate organism for a challenge study is typically a nonpathogenic organism that has similar survival capabilities and susceptibility to injury as the target pathogen and closely mimics how the pathogen would react under similar processing conditions [9,10]. A variety of organisms have been used as surrogates in place of pathogens to mimic pathogenic responses in commercial food processes, predominantly *E. faecium*, *Pediococcus* spp., and Biotype 1 *E. coli*. *Enterococcus faecium* ATCC 8459 (NRRL B-2354), used as a surrogate for *Salmonella* Enteritidis PT 30 in the thermal processing of wheat flour [11], as a *S. enterica* surrogate for storage time and temperature of milk powders [12], in thermal extrusion of low-moisture foods [13], and in plant-level validation of thermal processes for peanuts and pecans [14]. Investigators also found that *Pediococcus* strains had similar heat tolerances to *Salmonella* spp. and would be suitable surrogates for validation studies of jerky-style dried meat products [15–17]. *Pediococcus acidilactici* ATCC 8042 was examined as a *Salmonella* surrogate for thermal processing of toasted oats for cereal and peanuts for peanut butter [18], and for processing of low-moisture pet food [19]. Biotype 1 *E. coli* ATCC BAA-1427, BAA-1428, BAA-1429, and BAA-1430 have been used as thermal surrogates for *E. coli* O157:H7 in meat processes [20], as *Salmonella* surrogates for thermal processing of ground beef [21], and for thermal treatment of almonds and pistachios [22,23]. These strains have been recommended by USDA-FSIS as surrogate indicator organisms for food process validation studies [8].

Despite the prevalence of studies performed with surrogate bacteria for various food processes, no surrogate organisms have been proven to suitably represent the response of pathogens during biltong processing. The objective of this study was to examine potential nonpathogenic lactic acid bacteria and generic *E. coli* strains that could be used for in-plant studies to mimic pathogen lethality during biltong processing.

2. Materials and Methods

2.1. Bacterial Strains and Growth Conditions

Bacterial cultures used in this study were obtained from various sources including our laboratory culture collection, commercial starter cultures, and bacteria isolated from biltong trials as listed in Table 1.

Table 1. List of strains used as challenge organisms for biltong processing in this study.

Organism	Strain Designation	Culture Collection Designation	Antibiotic Resistance (μg/mL) *	Source
Pediococcus acidilactici	ATCC 8042	PMM 128	GM, 10; RF, 5	Muriana Culture Collection
Pediococcus acidilactici	PO2K5	PMM 331	GM, 10; RF, 5	Muriana Culture Collection
Pediococcus pentosaceous	ATCC 43200	PMM 104	GM, 10; RF, 5	Muriana Culture Collection
Pediococcus pentosaceous	FBB61-2	PMM 105	GM, 10; RF, 5	Muriana Culture Collection
Pediococcus acidilactici	Saga200	PMM 444	NA, 10; CL, 10	Kerry Foods, Beloit, WI, USA
Enterococcus faecium	201224-016	PMM 445	NA, 10; CL, 10	Vivolac Cultures, Indianapolis, IN, USA
Escherichia coli	ATCC BAA-1427	PMM 876	OX, 1; NB, 2.5	ATCC, Muriana Culture Collection
Escherichia coli	ATCC BAA-1428	PMM 877	OX, 1; NB, 2.5	ATCC, Muriana Culture Collection
Escherichia coli	ATCC BAA-1429	PMM 878	OX, 1; NB, 2.5	ATCC, Muriana Culture Collection
Escherichia coli	ATCC BAA-1430	PMM 879	OX, 1; NB, 2.5	ATCC, Muriana Culture Collection
Latilactobacillus sakei	GO-R2C	PMM 446	GM, 2.5; RF, 2.5	Isolated from biltong
Latilactobacillus sakei	GO-R2D	PMM 447	GM, 2.5; RF, 2.5	Isolated from biltong
Carnobacterium divergens	GO-R2E-B	PMM 448	GM, 2.5; RF, 2.5	Isolated from biltong
Carnobacterium divergens	GO-R1B	PMM 449	GM, 2.5; RF, 2.5	Isolated from biltong
Carnobacterium gallinarum	NB-R2A	PMM 450	GM, 2.5; RF, 2.5	Isolated from biltong
Carnobacterium gallinarum	NB-R2B	PMM 451	GM, 2.5; RF, 2.5	Isolated from biltong

* Antibiotic designations: gentamicin, GM; rifamycin, RF; nalidixic acid, NA; colistin, CL; oxacillin, OX; novobiocin, NB.

Bacterial isolates obtained from previous biltong beef trials after marination and drying for 8 days at 24.9 °C (75 °F) and 55% relative humidity (RH) were identified by 16S rRNA PCR/sequencing [24] as *Carnobacterium gallinarum*, *Carnobacterium divergens*, and *Latilactobacillus sakei* for examination as biltong process surrogates (Table 1).

Other lactic acid bacteria used in this study included *Pedicoccus acidilatcici* ATCC 8042, *P. acidilactici* P02K5, *P. pentosaceus* FBB61-2, and *P. pentosaceus* ATCC 43200, which are maintained in our laboratory culture collection. Some of these strains have been evaluated in other surrogate studies [19,25]. Nonpathogenic *E. coli* ATCC BAA-1427, BAA-1428, BAA-1429, and BAA-1430 have been used as Biotype 1 surrogate strains in various process validation studies and recommended for such use by USDA-FSIS [8,20,26]. *P. acidilactici* Saga200, used as a protective starter culture, was obtained as a frozen slurry from Kerry Foods (Beloit, WI, USA). *Enterococcus faecium* 201224-016 was obtained as a freeze-dried powder from Vivolac Cultures (Indianapolis, IN, USA) and is sold as a probiotic.

Carnobacterium spp., *E. faecium*, and *E. coli* cultures were inoculated into tryptic soy broth (TSB, BD Bacto, Franklin Laes, NJ, USA) and grown at 30 °C for 24 h. *L. sakei* and *Pedicoccus* spp. were inoculated into De Man, Rogosa and Sharpe broth (MRS, BD Bacto) and grown at 30 °C for 24 h. Cultures were prepared for storage by centrifugation (7200× *g*, 5 °C) of 9 mL of fresh, overnight culture, and the resulting pellet was resuspended with 2–3 mL of fresh, sterile TSB or MRS broth containing 10% glycerol. The cells in freezing media were then placed in 8 mL sterile glass vials and stored in an ultralow-temperature freezer (−80 °C) until use. Prior to use, frozen stocks were revived by transferring 100 μL of partially thawed culture into 9 mL of either TSB or MRS broth and incubated overnight at 30 °C.

Several cultures were used directly after suspension from the freeze-dried or frozen state for comparison of biltong process performance with metabolically active forms grown in liquid media. Prior to use, *P. acidilactici* Saga200 (frozen) was resuspended by adding 0.5 g of the frozen culture to 9 mL of 0.1% buffered peptone water (BPW, BD Difco) and vortexing until completely incorporated. *E. faecium* 201224-016 was resuspended by adding 0.1 g of the freeze-dried culture to 9 mL of 0.1% BPW and vortexing until completely mixed.

2.2. Acid Adpation of Cultures

Acid adaptation of active four-strain mixtures of *Carnobacterium* spp., *Pediococcus* spp., and *E. coli* BAA-strains was conducted as first described by Wilde et al. [27] and as used in previous biltong studies [4,28]. In brief, individual cultures were inoculated into TSB or MRS containing 1% glucose, incubated overnight at 30 °C, and harvested by centrifugation; cell pellets were then resuspended with 0.1% BPW. For mixed culture biltong inocula, individual strains were cultured, centrifuged, resuspended, and then combined in equal proportions to create a mixed inoculum cocktail. The commercial starter cultures (*P. acidilactici* Saga200 and *E. faecium* 201224-016) were not acid-adapted and used as a single-strain inoculum.

2.3. Beef Sample Preparation and Inoculation

USDA select-grade boneless beef rounds were obtained from a local meat processor (Ralph's Perkins, OK, USA) who obtains beef from a wholesale beef broker. Beef rounds were trimmed of fat and cut into approximately 5.1 cm wide × 1.9 cm thick × 7.6 cm long beef squares and held overnight at 5 °C on foil-lined trays wrapped in plastic bags. Beef pieces were inoculated the following morning with the respective inoculum depending on the trials being performed that day. Beef pieces were inoculated with the *Carnobacterium* spp. mixture (*C. divergens* GO-R2E-B, GO-R1B; *C. gallinarum* NB-R2A, NB-R2B), the *L. sakei* mixture (*L. sakei* GO-R2C, GO-R2D), the *Pediococcus* spp. mixture (*P. acidilactici* ATCC 8042, PO2K5; *P. pentosaceous* ATCC 43200, FBB61-2), *P. acidilactici* Saga200, or *E. faecium* 201224-016. The inoculum suspension (150 µL) was applied to each side of the beef pieces and immediately spread with a gloved finger. Inoculated beef pieces were then allowed to incubate for 30 min at 5 °C to allow for bacterial attachment prior to use.

2.4. Biltong Processing, Marination, and Drying

Biltong processing was conducted as described whereby trials were performed in duplicate and triplicate samples were harvested at each timepoint (n = 6) [4,29]. Following inoculation and attachment, the beef pieces were then dipped in sterile water to mimic rinse treatments that processors often apply using antimicrobials or water during processing. The inoculated pieces were placed in a plastic basket, dipped in sterile water in a stainless-steel tub for 30 s, and drained for 60 s to release excess liquid. The beef pieces were then placed into a chilled metal tumbling vessel containing a biltong marinade. The biltong marinade consisted of 2.2% salt, 0.8% black pepper, 1.1% coarse ground coriander, and 4% red wine vinegar (100-grain; 10% acetic acid) in relation to the total meat weight. Beef pieces were vacuum-tumbled (15 inches Hg) in a Biro VTS-43 vacuum-tumbler (Marblehead, OH, USA) for 30 min and then hung to dry in a humidity-controlled oven (Hotpack, Model 435315, Warminster, PA, USA) at 55% relative humidity and 24.9 °C (75 °F) for 8–10 days.

2.5. Selective Recovery of Inoculum Bacteria from Biltong-Inoculated Beef

The bacteria assessed in this study as potential biltong processing surrogates were inoculated onto raw beef, and initial and residual inoculum enumeration had to preclude other natural contaminants also found on raw beef, those contributed during trimming of beef, or from the marinade spice mix. Prior studies indicated that such processing conditions induce stresses, and injured cells may not be recovered on harsh selective media, thereby giving a falsely lower count [28]. To eliminate the possibility of inhibiting injured-but-viable cells, we used generic growth media (TSA, MRS agar) supplemented with antibiotics to which the strains are resistant as a selective medium to enumerate our inoculated organisms from samples taken during biltong processing [4,28]. Antibiotic resistance was determined using antibiotic susceptibility discs (BD BBL Sensi-Discs, BD Labs, Franklin Lakes, NJ, USA) to determine innate antibiotic resistance (Table 1). After identification of antibiotic resistances, cultures were then enumerated on media with and without antibiotics to ensure the absence of inhibition from the use of antibiotics in the media as described previously [4,28,30]. For some strains used as inoculum cocktails that did not have consensus of the same antibiotic resistances, antibiotic resistance was acquired by plating on low level antibiotics known to generate spontaneous antibiotic resistance (i.e., gentamycin and rifamycin).

2.6. Comparison of Commerically Available Starter Cultures as Biltong Inoculants in their Lyophilized and Metabolically Active Forms

2.6.1. Culture Preparation

Lactic acid bacteria obtained as freeze-dried cultures from starter culture companies for use in validation studies may present a facile method of use as validation inocula by simply resuspending the cells in buffer and directly inoculating beef samples [15,17]. Freeze-drying or lyophilization of bacteria exposes them to stressful conditions that can affect subsequent cell viability or activity [20,21]. Therefore, the activity of lyophilized (*E. faecium* 201224-016) and frozen (*P. acidilactici* Saga200) starter cultures and their metabolically active forms (i.e., after growth in media) were compared in their response to biltong processing.

For the lyophilized culture (*E. faecium* 201224-016), 0.1 g of freeze-dried powder was added to 9 mL of sterile 0.1% BPW and vortexed until completed suspended. The resuspended mixture was then used to inoculate each beef piece (300 μL; 150 μL/side) prior to marination.

For the frozen starter culture (*P. acidilactici* Saga200), a sterile hollow hole puncher was used to core ~0.8 g of frozen Saga200 from the manufacture's container which was added to 9 mL of sterile 0.1% BPW and vortexed until mixed. The culture suspension was kept chilled on ice and used shortly thereafter to inoculate beef pieces.

Metabolically active versions of these cultures were obtained by growth in 150 mL of the appropriate media (TSB, MRS) for 24 h at 30 °C, centrifugation, and resuspension of the recovered cell pellet with 5 mL of sterile 0.1% BPW. The resuspended culture was then used to inoculate beef pieces prior to use in the validation study. The lyophilized and metabolically active forms of *E. faecium* 201224-016 and *P. acidilactici* Saga200 were used in parallel and simultaneous biltong trials to reduce any variables that might influence the observed effect of the marinade and drying process.

2.6.2. Lyophilization of *Carnobacterium gallinarum* NB-R2A

To the authors' knowledge, there is no commercially available *Carnobacterium* strain available in the United States. Therefore, *C. gallinarum* NB-R2A, isolated from biltong, was lyophilized via freeze-drying to examine a lyophilized version for comparison with the actively grown culture. *Carnobacterium gallinarum* NB-R2A was inoculated into 9 mL of TSB from frozen stock and incubated for 18 h at 30 °C. Following incubation, the 9 mL culture was transferred to 190 mL of TSB and incubated again for 18 h at 30 °C. The culture was then centrifuged at 7200× *g* for 20 min. The supernatant was removed, and the cell pellet was resuspended with 5 mL of sterile BPW and repeated. The supernatant was removed following centrifugation, and the final cell pellet was resuspended with 10 mL of autoclaved milk-based freeze-drying medium consisting of 11 g of skim milk powder, 1 g of dextrose, 1 g of trehalose, and 0.2 g of yeast extract per 100 mL. The milk/cell suspension was added to Oak Ridge tubes (5 mL each) and freeze-dried using a Heto vacuum centrifuge (Model VR-maxi) connected to a Heto freezing condensor (Model CT 60E) and a Leybold Trivac vacuum pump (Model D2.5F) setup for 24 h under vacuum. The freeze-dried powder was then stored at −80 °C until use in our biltong study. Just before use, 0.25 g of powder was added to 9 mL of sterile 0.1% BPW, vortexed until mixed, and used to inoculate beef pieces for biltong processing.

2.7. Evaluation of Critical Parameters and Intrinsic Factors in the Biltong Process

2.7.1. Water Activity

Uninoculated beef pieces were sampled for water activity (A_w) measurements at various stages throughout processing (in triplicate) including the initial raw beef, beef after marination, and then beef after drying for 2, 4, 6, 8, and 10 days. To obtain measurements, beef pieces were cut in half and placed in a sampling cup with the interior portion of the sample facing upward (toward the sensor). Samples were then covered with sampling cup cover containing the sensor and allowed to equilibrate to the temperature of the room. Water activity was measured using a HC2-AW-USB probe with a direct PC interface and HW4-P-Quick software (Rotronic Corp., Hauppauge, NY, USA). Measurements were taken in triplicate for each sample at each timepoint.

2.7.2. Moisture Loss

Following marination, each beef piece was individually weighed and labeled prior to being hung in the humidity-controlled oven. Three pieces were selected and weighed prior to processing, and then sampled every 2 days while drying. The weight at the time of sampling was compared to the initial weight of the same piece recorded prior to drying. The determination of percent moisture loss was calculated as per Equation (1).

$$\%\ Moisure\ Loss = \frac{[(inital\ weight) - (final\ weight)]}{(inital\ weight)} \times 100 \quad (1)$$

2.7.3. Measurement of Biltong Beef pH

Measurements of beef pH were obtained at various points in the biltong process including raw beef, beef following marination, and beef after 2, 4, 6, 8, and 10 days of drying. At each timepoint, three pieces of uninoculated beef were collected, weighed, and then added to a laboratory blender with steel blades (Waring Commercial, New Harford, CT, USA) with sterile water of equal weight to the weight of the beef pieces. The water and beef mixtures were blended until a finely ground mixture was formed. The pH of the homogenized meat mixture was measured in triplicate using an H-series pH meter and probe (Hach, Loveland, CO, USA).

2.7.4. Salt Concentration

The homogenized meat mixture used to measure pH was also used to obtain salt concentrations of each sample. Horiba LAQUA Twin Pocket Meter (Horiba Instruments, Irvine, CA, USA) was used to quantify sodium ion concentration. Approximately 300 µL of the homogenized sample was placed in the sample chamber and allowed to stabilize before recording. Readings (in ppm) were taken in triplicate for each sample. To determine the salt (*NaCl*) concentration from the sodium ion concentration, the following equations were used:

$$Na\left(\frac{\text{mg}}{100\ \text{g}}\right) = Meter\ reading\ (\text{ppm}) \times \frac{Weight\ after\ Dilution\ (\text{g})}{Sample\ Weight} \times 100, \quad (2)$$

$$NaCl\ Salt\left(\frac{\text{g}}{100\ \text{g}}\right) = Na\left(\frac{\text{mg}}{100\ \text{g}}\right) \times \frac{NaCl\ molar\ mass}{Na\ molar\ mass} \times \frac{1}{1000}. \quad (3)$$

2.8. *Microbial Sampling and Inoculum Enumeration of Biltong Beef*

At each sampling timepoint of biltong beef processing (raw beef, after marinade, and after every 2, 4, 6, 8, and 10 days of drying), three beef pieces were selected at random and placed in a sterile Whirl-pak filter stomaching bag (Nasco, Fort Atkinson, WI, USA) in combination with 100 mL of 1% neutralizing buffered peptone water (nBPW, Criterion, Hardy Diagnostics, Santa Maria, CA, USA). Samples were stomached for 60 s in a paddle-blender masticator (IUL Instruments, Barcelona, Spain). Serial dilutions were made with 1% BPW and plated on TSA containing gentamicin and rifamycin (2.5 µg/mL each) for *Carnobacterium*, on MRSA containing gentamicin and rifamycin (2.5 µg/mL each) for *L. sakei*, on MRSA containing gentamicin (10 µg/mL) and rifamycin (5 µg/mL) for *Pediococcus* spp., on TSA containing naldixic acid and colistin (10 µg/mL each) for *E. faecium* 201224-016, or on MRS containing nalidixic acid and colistin (10 µg/mL each) for *P. acidilactici* Saga200; the filter bag dilution was considered the 10^{0} dilution. Plates were incubated at 30 °C for

48 h and enumerated as log CFU/mL. Samples were collected in triplicate replication and plated in duplicate at each sampling timepoint.

2.9. Statistical Analysis

Validation trials were conducted in duplicate with triplicate sampling at each timepoint ($n = 6$) as per validation criteria established by the National Advisory Committee on Microbial Criteria for Foods (NACMCF) [9] and supported by the USDA-FSIS [31]. Data are presented as the mean of multiple replications with standard deviation of the mean represented by error bars. Statistical analysis of data collected over time was performed using one-way repeated-measures analysis of variance (RM-ANOVA). Pairwise multiple comparisons were performed using the Holm–Sidak test to determine significant differences. Data treatments with the same letter are not significantly different ($p > 0.05$); treatments with different letters are significantly different ($p < 0.05$).

3. Results and Discussion

3.1. Critical Parameters and Intrinsic Factors

3.1.1. Water Activity, Moisture Loss, and Salt Concentrations

To complement the surrogate validation trials, we measured and recorded critical operational parameters and intrinsic factors at each key stage of processing (raw beef, inoculation, marination, and every 2 days of drying) as recommended by USDA-FSIS [1]. Water activity (A_W) is a measure of free, unbound water available for bacterial growth. USDA-FSIS considers vacuum-tumbled beef as 'nonintact beef', whereby A_W is a primary safety factor as there is no heat lethality step in biltong processing and biltong is processed as thick beef samples [32,33]. Therefore, A_W is a critical safety factor for control of bacteria that might be internalized due to vacuum tumbling. *S. aureus* that can tolerate low A_W and high salt levels would be a concern for possible production of staphylococcal enterotoxin. The targeted A_W for shelf-stable beef jerky is <0.85 which was achieved after 7 days of drying (Figure 1) [1,2]. Water activity after 8 and 10 days of drying ranged from 0.82 to 0.79 respectively. Similarly, beef samples showed incremental moisture loss with 59% and 62.5% loss at 8 and 10 days, respectively (Figure 1).

Salt concentration was also determined during the biltong process. Salt concentration was calculated from sodium readings obtained with the LAQUAtwin NA-11 sodium ion meter (Horiba Inc, Irvine, CA, USA). The initial calculated salt concentration determined on raw beef was 0.12% NaCl; then, following the marination step, the beef salt concentration shot up to 2.17% (2.17 g NaCl/100 g beef). The initial salt level falls in line with expectations given that the biltong marinade was formulated at 2.2% salt (*w*/*w*). The salt concentration increased over time and was indirectly proportional to moisture loss during the drying process (Figure 2). As expected, as moisture loss occurred, A_W was also reduced to below 0.85 A_W (Figure 1) and the salt concentration increased to above 4% (Figure 2); both conditions are inhibitory to most bacteria, helping to ensure a safe product for consumers [34]. Biltong safety involves an interplay among moisture, salt concentration, and A_W since moisture loss increases salt concentration, while salt binds water and helps to draw it out of the interior of the beef, thereby reducing A_W. For consumer issues regarding high sodium levels, the use of alternative salts ($CaCl_2$, KCl) instead of NaCl can help lower sodium levels in finished biltong while still maintaining a 5 log reduction of pathogen (*Salmonella*) [29].

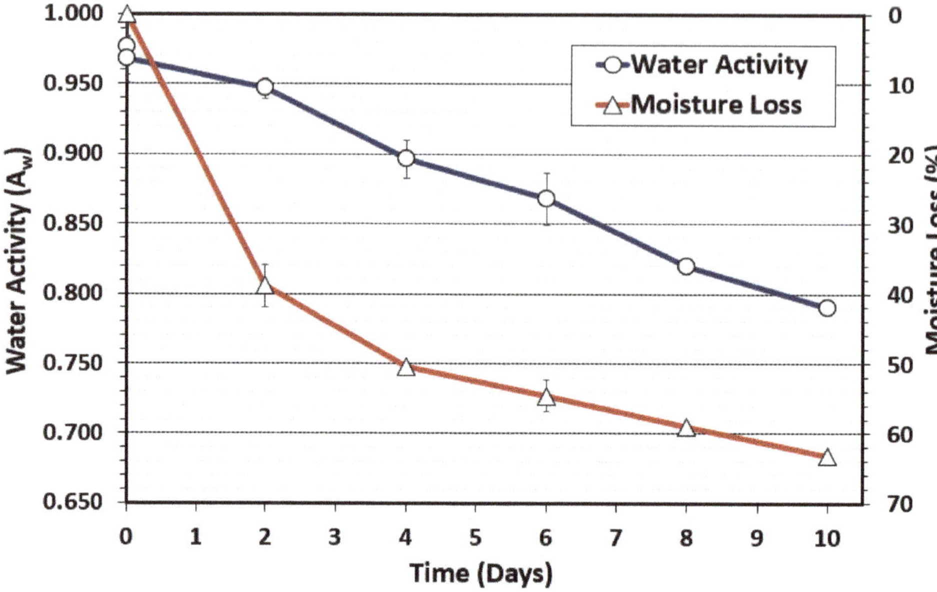

Figure 1. Water activity (A_W) and moisture loss during biltong processing at 24.9 °C (75 °F) and 55% RH. The data represent the average of measurements taken during duplicate trials with triplicate samples taken at each time interval (n = 6).

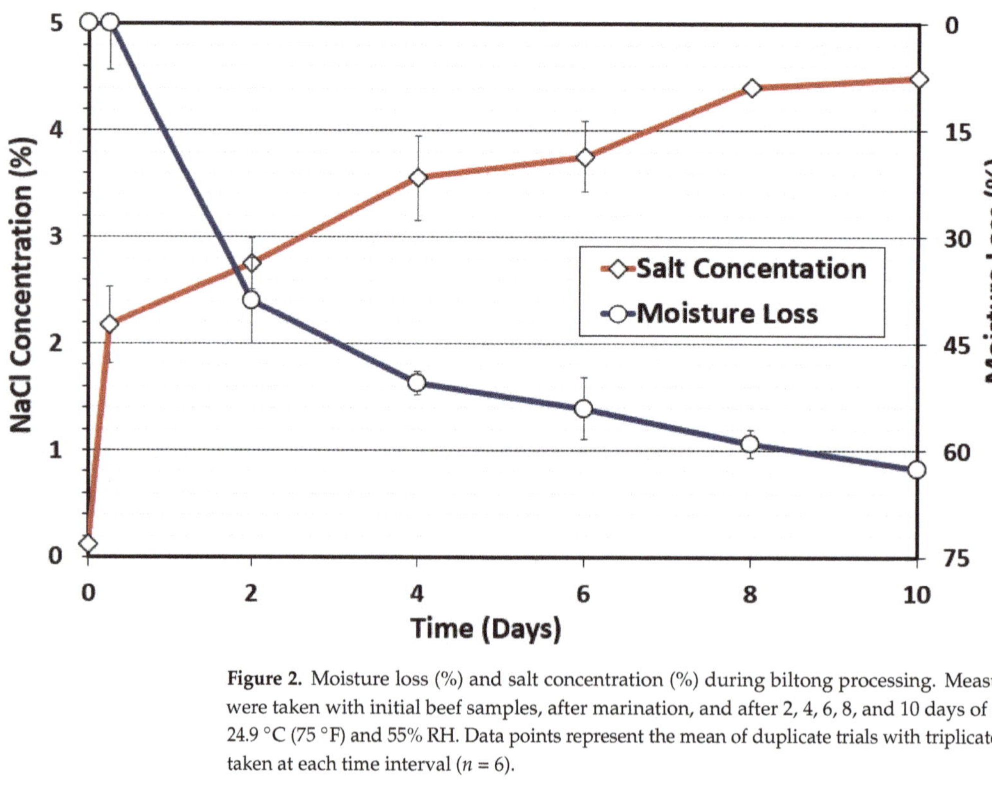

Figure 2. Moisture loss (%) and salt concentration (%) during biltong processing. Measurements were taken with initial beef samples, after marination, and after 2, 4, 6, 8, and 10 days of drying at 24.9 °C (75 °F) and 55% RH. Data points represent the mean of duplicate trials with triplicate samples taken at each time interval (n = 6).

3.1.2. The pH of Beef during Biltong Processing

The initial pH of the raw meat pieces was approximately pH 5.43 (Figure 3), which was determined by blending beef samples in sterile water in a laboratory blender. The pH of the samples then decreased following the marination step to 5.02, which can be attributed to the presence of residual 100-grain red wine vinegar in the marinade. After removal from the marinade, the pH of biltong beef samples then equilibrated slightly higher to ~5.18–5.20 for the remainder of the drying process in the humidity-controlled oven (Figure 3). The pH of the marination solution was much lower (pH 2.5–2.7); during 30 min vacuum tumbling, the surface bacteria were immersed in the low-pH marinade solution, which could lead to cell death and inactivation of pathogenic bacteria [35,36], as observed in the current study and prior biltong trials where levels of inoculated pathogens were reduced after marination [4]. After removal from the vacuum tumbler, the residual marinade on the surface was absorbed, and the pH of biltong beef samples equilibrated to ~5.18–5.20 for the remainder of the drying process in the humidity oven (Figure 3).

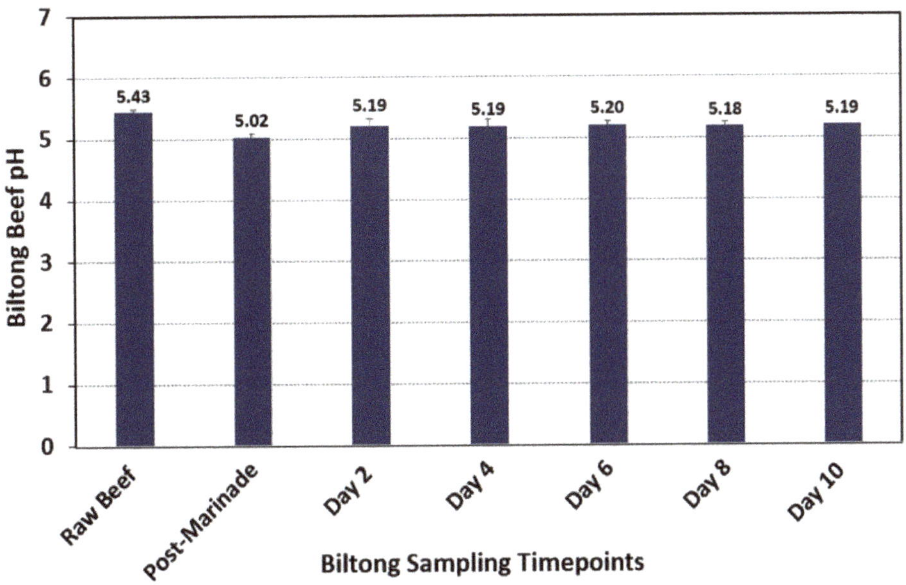

Figure 3. The pH of meat at each sampling timepoint during biltong processing. Samples were taken in triplicate at each timepoint following blending with sterile water in a laboratory blender ($n = 6$).

3.1.3. Temperature and Relative Humidity during Biltong Processing

Temperature and RH measurements were recorded by computer software connected to the handheld temperature and humidity recorders to which the probes in the oven chamber were connected (Figure 4). Two temperature probes were inserted separately into two beef pieces to measure the internal beef temperature during processing, while the humidity probe was placed midway within the chamber. Air temperature and humidity were set to 23.9 °C (75 °F) and 55% throughout the duration of each trial but cycled above and below the set points. The internal temperature of the beef was more consistent and steadily increased from their initial temperature to match the temperature of the chamber. Long-term storage at low RH helps to evaporate moisture from the beef.

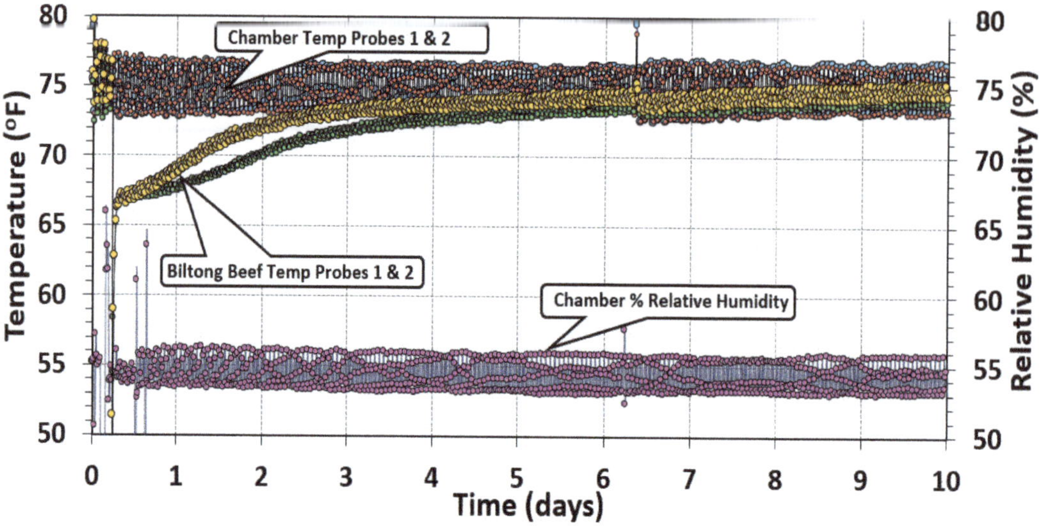

Figure 4. Oven temperature and relative humidity measurements. The temperature was set to 24.9 °C (75 °F), and the relative humidity setpoint was 55% RH during the drying process over a period of 10 days. Graphical data show the typical cycling of oven control above/below setpoint. Two temperature probes were placed in various places in the chamber and two additional probes were inserted into separate pieces of beef to track the internal temperature of the biltong product over the same drying period.

3.2. Surrogate Log Reductions during Biltong Processing

Various bacteria were considered for examination as possible nonpathogenic surrogates, including strains recovered from biltong after processing. These included a two-strain mixture of *L. sakei* GO-R2C and GO-R2D and a four-strain mixture of *C. divergens* GO-R2E-B and GO-R1B and *C. gallinarum* NB-R2A and NB-R2B (Figure 5). We also examined a four-strain mixture of *P. acidilactici* and *P. pentosaceous* strains (*P. acidilactici* ATCC 8042 and PO2K5; *P. pentosaceous* ATCC 43200 and FBB61-2) vs. starter cultures that were available through culture companies (*E. faecium* 201224-016 and *P. acidilactici* Saga200) as surrogate organisms (Figure 5).

Only a slight reduction from inoculated levels was observed following vinegar/spice/salt marination (0.65, 0.58, 0.75, and 0.61 log reduction) with all cultures used, except for the four-strain mixtures of *Carnobacterium* spp. and *E. coli* ATCC BAA series (Figure 5). A larger log reduction was observed after marination of the four-strain mixtures of *Carnobacterium* spp. (1.23 log) and *E. coli* ATCC BAA-strains (0.86 log) (Figure 5). Trials using *E. coli* ATCC BAA (four-strain mix), *L. sakei* (two-strain mix), *Pediococcus* spp. (four-strain mix), *E. faecium* 201224-016, and *P. acidilactici* Saga200 failed to achieve a 5 log reduction during biltong processing with overall reductions of 4.86 log, 2.03 log, 1.87 log, 1.68 log, and 1.83 log respectively. Of all the nonpathogenic strains examined, only the four-strain mixture of *Carnobacterium* spp. achieved an overall reduction of greater than 5 log (5.85 log) during the 8 day drying period (Figure 5). On the basis of these results, *Carnobacterium* spp. were the only organisms that achieved a 5 log reduction (within 6–8 days) comparable to that observed for the pathogenic strains, and they presented the best case for use as a *Salmonella*, *L. monocytogenes*, *E. coli* O157:H7, or *S. aureus* surrogate for biltong processing (Figure 5).

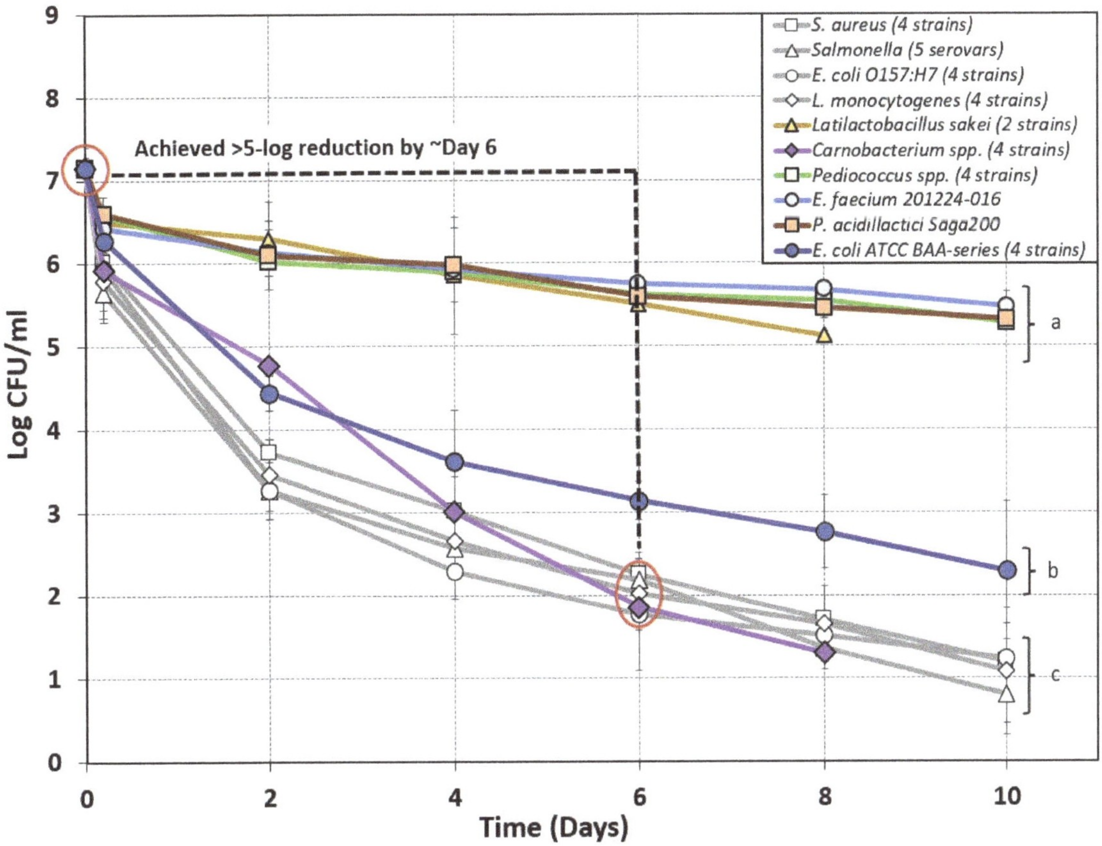

Figure 5. Composite graph of biltong processing data of nonpathogenic bacteria attempting to mimic the biltong process log reduction of pathogenic bacteria (light-gray lines) to be considered a possible 'biltong processing surrogate' organism for in-plant validation. Log reduction curves of various lactic acid bacteria (*Carnobacterium* spp., *Pediococcus* spp., *L. sakei*, and *E. faecium*) and Biotype I *E. coli* strains tested as potential surrogate organisms for biltong processing over a period of 8–10 days. Strains were compared to the log reduction curves observed during previous biltong validation studies using pathogenic bacteria including *Salmonella* serovars [4], *S. aureus*, *E. coli* O157:H7, and *L. monocytogenes* [5]. Data points are the mean of duplicate trials sampled in triplicate ($n = 6$). Statistical analysis was performed using one-way repeated-measures analysis of variance (RM-ANOVA) of the entire time course of data; curves with the same letter are not significantly different ($p > 0.05$); isolates with different letters are significantly different ($p < 0.05$).

3.3. Comparison of Lyophilized/Frozen Starter Cultures with Metabolically Active (Grown) Versions in Biltong Processing Trials

Several reports in the literature have used freeze-dried or frozen cultures, resuspended directly in buffer, to inoculate food samples in process trials for direct comparison to pathogens grown in microbiological media (which we describe as 'active cultures') [15,17]. The ease of availability of freeze-dried/frozen cultures from culture companies would facilitate the use of such cultures for in-plant validation studies; however, we were interested to see if they could provide the same response in a biltong process as the actively grown cultures (Figure 6). The comparisons were between two commercially available starter cultures, *E. faecium* 201224-016 (Vivolac Cultures; freeze-dried) and *P. acidilactici* Saga200 (Kerry Foods; frozen), and a lyophilized *C. divergens* NB R2A, which was chosen from among the *Carnobacterium* mixed strains demonstrating >5 log reduction in Figure 5.

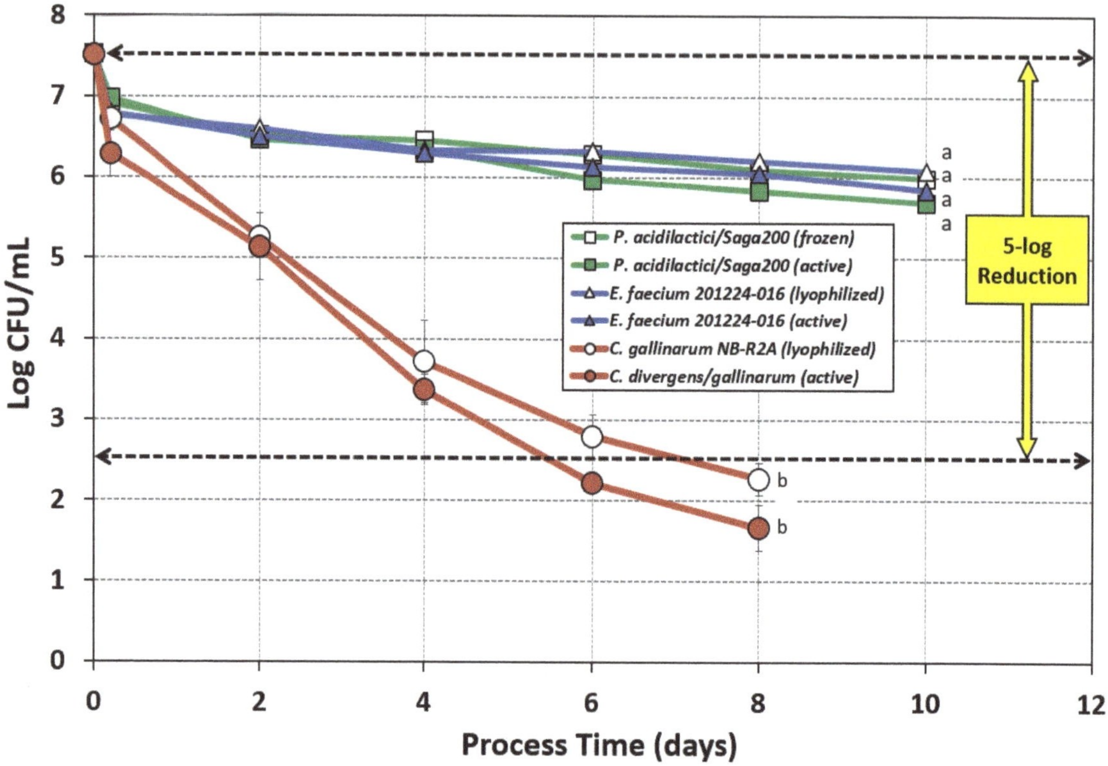

Figure 6. Biltong processing of beef inoculated with lyophilized/frozen cells vs. metabolically active cells (freshly grown) of *E. faecium* 201224-016, *P. acidilactici*, and *C. gallinarum* NB-R2A. Lyophilized *C. gallinarum* NB-R2A was compared to a four-strain cocktail of metabolically active *C. divergens/gallinarum*. Graph curves of frozen or lyophilized cultures have hollow symbols. Statistical analysis was performed using one-way repeated-measures analysis of variance (RM-ANOVA) over the entire time course of the datasets; graphs with the same letter are not significantly different ($p > 0.05$); isolates with different letters are significantly different ($p < 0.05$).

Neither the lyophilized version of *E. faecium* 201224-016 (1.43 log reduction) nor the frozen version of *P. acidilactici* Saga200 (1.54 log reduction) achieved the 5 log reduction target; survival curves of the lyophilized/frozen forms were also not significantly different when compared to their metabolically active forms, i.e., 1.68 and 1.83 log reduction, respectively (Figure 6). The lyophilized single strain *C. divergens* NB R2A also showed no significant difference from the metabolically active culture and again achieved 5 log reduction during the biltong process (Figure 6). The data show that lyophilized or frozen versions of *E. faecium, P. acidilactici*, or *C. gallinarum* do not respond differently than actively grown cultures to biltong processing conditions and, when possible, their use might facilitate inoculated studies.

4. Conclusions

The lethality observed in the biltong process with *Carnobacterium* spp. aligned with that observed with four major pathogenic organisms indicating that *Carnobacterium* spp. could be an effective in-plant surrogate organism to monitor the effectiveness of biltong processing within a manufacturer's facility. *Enterococcus faecium, L. sakei*, and *Pediococcus* spp. were not reduced much (<2 log) and were resilient toward the acid, salt, and low A_w experienced during 10 days of biltong processing. The use of lyophilized/frozen cells as inoculum for validation of biltong processing was not significantly different than using actively grown cells. This work helps to fill USDA-FSIS knowledge gaps in air-dried shelf-stable dried beef (biltong) processing with regard to potential surrogate organisms and critical factors involved in the biltong process. Future studies on biltong processing may include whether pathogens such as *Salmonella*, known to survive long periods of low water activity, can survive the extended shelf-life of biltong products to ensure that this does not become a possible (overlooked) problem.

Author Contributions: Conceptualization, P.M.M.; methodology, P.M.M. and C.E.K.; software, P.M.M.; validation, C.E.K. and P.M.M.; formal analysis, C.E.K. and J.W.; investigation, C.E.K. and J.W.; resources, P.M.M.; data curation, P.M.M.; writing—original draft preparation, C.E.K.; writing—review and editing, P.M.M.; visualization, P.M.M.; supervision, P.M.M.; project administration, P.M.M.; funding acquisition, P.M.M. All authors have read and agreed to the published version of the manuscript.

Funding: Funding for the study was contributed in part by the Robert M. Kerr Food & Ag Products Center, the Advance Foods—SE Gilliland Professorship in Microbial Food Safety (#21-57200), the USDA National Institute of Food and Agriculture (Hatch Project #OKL03090), the Oklahoma Agricultural Experiment Station, the Division of Agricultural Sciences and Natural Resources at Oklahoma State University, and various industry sources.

Institutional Review Board Statement: The work performed in this study was performed under Protocol #17-8 approved by the OSU Institutional Biosafety Committee within the Office of University Research Compliance.

Informed Consent Statement: Not applicable.

Data Availability Statement: Not applicable.

Acknowledgments: The authors would like to acknowledge the 'mentoring team' supporting Caitlin Karolenko's USDA-FSIS 2021–2022 Graduate Fellowship Award, including Emilio Esteban (USDA-FSIS), Isabel Walls (USDA-FSIS; Washington, DC, USA), Meryl Silverman (USDA-FSIS; OPPD, Washington, DC, USA), Kristina Barlow (USDA-FSIS; Risk Management and Innovations Staff, OPPD, Washington, DC, USA), Terry Campbell (USDA-FSIS; OPHS, Eastern Laboratory, Athens, GA), and Steve W. Mamber (USDA-FSIS; OPARM, Washington, DC, USA), for their technical assistance in meat processing and USDA-FSIS regulatory policy with respect to dried beef.

Conflicts of Interest: The authors declare no conflict of interest.

References

1. USDA-FSIS. *FSIS Compliance Guideline for Meat and Poultry Jerky Produced by Small and Very Small Establishments*; U.S. Food Safety and Inspection Service: Washington, DC, USA, 2014; pp. 1–54.
2. USDA-FSIS. *Salmonella Compliance Guidelines for Small and Very Small Meat and Poultry Establishments That Produce Ready-to-Eat (RTE) Products and Revised Appendix A, June 2017*; U.S. Food Safety and Inspection Service: Washington, DC, USA, 2017; pp. 1–37.
3. Nickelson, R.; Luchansky, J.B.; Kaspar, C.; Johnson, E. An executive summary prepared by The Blue Ribbon Task Force of the National Cattlemen's Beef Association. In *Dry Fermented Sausage and Escherichia coli O157:H7 Validation Research*; Research Report No. 11-316; National Cattlemen's Beef Association: Chicago, IL, USA, 1996.
4. Karolenko, C.E.; Bhusal, A.; Nelson, J.L.; Muriana, P.M. Processing of biltong (dried beef) to achieve USDA-FSIS 5-log reduction of Salmonella without a heat lethality step. *Microorganisms* **2020**, *8*, 791. [CrossRef]
5. Gavai, K.; Karolenko, C.; Muriana, P.M. Effect of biltong dried beef processing on the reduction of *Listeria monocytogenes*, *E. coli* O157:H7, and *Staphylococcus aureus*, and the contribution of the major marinade components. *Microorganisms* **2022**, *10*, 1308. [CrossRef] [PubMed]
6. USDA-FSIS. Guidelines for preparing and submitting experimental protocols for in-plant trials of new technologies and procedures. *Fed. Reg.* **1995**, *60*, 2.
7. USDA-FSIS. FSIS procedures for notification of new technology. *Fed. Reg.* **2003**, *68*, 3.
8. USDA-FSIS. Use of Non-Pathogenic *Escherichia coli* (*E. coli*) Cultures as Surrogate Indicator Organisms in Validation Studies. Available online: https://ask.usda.gov/s/article/Use-of-Non-pathogenic-Escherichia-coli-E-coli-Cultures-as-Surrogate-Indicator-Organisms-in-Validation-Studies#:~{}:text=Can%20establishments%20use%20non-pathogenic%20E.%20coli%20cultures%20as,establishment%20to%20measure%20changes%20in%20microbial%20counts%3F%20Yes (accessed on 13 June 2022).
9. National Advisory Committee on Microbiological Criteria for Foods. Parameters for determining inoculated pack/challenge study protocols. *J. Food Prot.* **2010**, *73*, 140–202. [CrossRef] [PubMed]

10. Hu, M.; Gurtler, J.B. Selection of surrogate bacteria for use in food safety challenge studies: A review. *J. Food Prot.* **2017**, *80* 1506–1536. [CrossRef]
11. Liu, S.; Rojas, R.V.; Gray, P.; Zhu, M.-J.; Tang, J. *Enterococcus faecium* as a Salmonella surrogate in the thermal processing of wheat flour: Influence of water activity at high temperatures. *Food Microbiol.* **2018**, *74*, 92–99. [CrossRef]
12. Wei, X.; Agarwal, S.; Subbiah, J. Evaluation of *Enterococcus faecium* NRRL B-2354 as a surrogate for Salmonella enterica in milk powders at different storage times and temperatures. *J. Dairy Sci.* **2021**, *104*, 198–210. [CrossRef]
13. Bianchini, A.; Stratton, J.; Weier, S.; Hartter, T.; Plattner, B.; Rokey, G.; Hertzel, G.; Gompa, L.; Martinez, B.; Eskridge, K.M. Use of *Enterococcus faecium* as a surrogate for Salmonella enterica during extrusion of a balanced carbohydrate-protein meal. *J. Food Prot.* **2014**, *77*, 75–82. [CrossRef]
14. Brar, P.K.; Danyluk, M.D. Validation of *Enterococcus faecium* as a surrogate for Salmonella under different processing conditions for peanuts and pecans. *Food Microbiol.* **2019**, *80*, 9–17. [CrossRef] [PubMed]
15. Borowski, A.G.; Ingham, S.C.; Ingham, B.H. Validation of ground-and-formed beef jerky processes using commercial lactic acid bacteria starter cultures as pathogen surrogates. *J. Food Prot.* **2009**, *72*, 1234–1247. [CrossRef] [PubMed]
16. Dierschke, S.; Ingham, S.C.; Ingham, B.H. Destruction of *Escherichia coli* O157:H7, *Salmonella*, *Listeria monocytogenes*, and *Staphylococcus aureus* achieved during manufacture of whole-muscle beef jerky in home-style dehydrators. *J. Food Prot.* **2010**, *73*, 2034–2042. [CrossRef]
17. Borowski, A.G.; Ingham, S.C.; Ingham, B.H. Lethality of home-style dehydrator processes against *Escherichia coli* O157:H7 and Salmonella serovars in the manufacture of ground-and-formed beef jerky and the potential for using a pathogen surrogate in process validation. *J. Food Prot.* **2009**, *72*, 2056–2064. [CrossRef] [PubMed]
18. Deen, B.; Diez-Gonzalez, F. Assessment of *Pediococcus acidilactici* ATCC 8042 as potential *Salmonella* surrogate for thermal treatments of toasted oats cereal and peanut butter. *Food Microbiol.* **2019**, *83*, 187–192. [CrossRef]
19. Ceylan, E.; Bautista, D.A. Evaluating *Pediococcus acidilactici* and *Enterococcus faecium* NRRL B-2354 as thermal surrogate microorganisms for Salmonella for in-plant validation studies of low-moisture pet food products. *J. Food Prot.* **2015**, *78*, 934–939. [CrossRef]
20. Keeling, C.; Niebuhr, S.E.; Acuff, G.R.; Dickson, J.S. Evaluation of *Escherichia coli* Biotype I as a Surrogate for *Escherichia coli* O157:H7 for Cooking, Fermentation, Freezing, and Refrigerated Storage in Meat Processes. *J. Food Prot.* **2009**, *72*, 728–732. [CrossRef]
21. Redemann, M.A.; Brar, J.; Niebuhr, S.E.; Lucia, L.M.; Acuff, G.R.; Dickson, J.S.; Singh, M. Evaluation of thermal process lethality for non-pathogenic *Escherichia coli* as a surrogate for Salmonella in ground beef. *LWT* **2018**, *90*, 290–296. [CrossRef]
22. Ma, L.; Kornacki, J.L.; Zhang, G.; Lin, C.M.; Doyle, M.P. Development of thermal surrogate microorganisms in ground beef for in-plant critical control point validation studies. *J. Food Prot.* **2007**, *70*, 952–957. [CrossRef]
23. Almond Board of California. *Guidelines for Process Validation Using Enterococcus faecium NRRL B-2354*; Almond Board of California: Modesto, CA, USA, 2007.
24. Shah, K.; Muriana, P.M. Efficacy of a next generation quaternary ammonium chloride sanitizer on *Staphylococcus* and *Pseudomonas* biofilms and practical application in a food processing environment. *Appl. Microbiol.* **2021**, *1*, 89–103. [CrossRef]
25. de Souza de Azevedo, P.O.; de Azevedo, H.F.; Figueroa, E.; Converti, A.; Domínguez, J.M.; de Souza Oliveira, R.P. Effects of pH and sugar supplements on bacteriocin-like inhibitory substance production by *Pediococcus pentosaceus*. *Molec. Biol. Rep.* **2019**, *46*, 4883–4891. [CrossRef]
26. Niebuhr, S.E.; Laury, A.; Acuff, G.R.; Dickson, J.S. Evaluation of Nonpathogenic Surrogate Bacteria as Process Validation Indicators for Salmonella enterica for Selected Antimicrobial Treatments, Cold Storage, and Fermentation in Meat. *J. Food Prot.* **2008**, *71*, 714–718. [CrossRef]
27. Wilde, S.; Jørgensen, F.; Campbell, A.; Rowbury, R.; Humphrey, T. Growth of Salmonella enterica Serovar Enteritidis PT4 in media containing glucose results in enhanced RpoS-independent heat and acid tolerance but does not affect the ability to survive air-drying on surfaces. *Food Microbiol.* **2000**, *17*, 679–686. [CrossRef]
28. Karolenko, C.E.; Bhusal, A.; Gautam, D.; Muriana, P.M. Selenite cystine agar for enumeration of inoculated Salmonella serovars recovered from stressful conditions during antimicrobial validation studies. *Microorganisms* **2020**, *8*, 338. [CrossRef] [PubMed]
29. Karolenko, C.; Muriana, P. Quantification of process lethality (5-Log reduction) of Salmonella and salt concentration during sodium replacement in biltong marinade. *Foods* **2020**, *9*, 1570. [CrossRef]
30. Bhusal, A.; Nelson, J.; Pletcher, D.; Muriana, P.M. Comparison of sodium nitrite and 'natural' nitrite on the inhibition of spore germination and outgrowth of *Clostridium sporogenes* in low- and high-fat frankfurters. *Appl. Microbiol.* **2021**, *1*, 104–122. [CrossRef]
31. USDA-FSIS (Ed.) *FSIS Compliance Guideline HACCP Systems Validation (April 2015)*; USDA-FSIS: Washington, DC, USA, 2015; pp. 1–68.
32. Pokharel, S.; Brooks, J.C.; Martin, J.N.; Brashears, M.M. Antimicrobial susceptibility and internalization of Salmonella Typhimurium in vacuum-tumbled marinated beef products. *Lett. Appl. Microbiol.* **2016**, *63*, 412–418. [CrossRef]
33. USDA-FSIS. Non-Intact Beef Products. Available online: https://ask.usda.gov/s/article/Non-intact-beef-products (accessed on 16 June 2022).
34. Gurtler, J.B.; Keller, S.E.; Kornacki, J.L.; Annous, B.A.; Jin, T.; Fan, X. Challenges in recovering foodborne pathogens from low-water-activity foods. *J. Food Prot.* **2019**, *82*, 988–996. [CrossRef]

35. Lund, P.A.; De Biase, D.; Liran, O.; Scheler, O.; Mira, N.P.; Cetecioglu, Z.; Fernández, E.N.; Bover-Cid, S.; Hall, R.; Sauer, M.; et al. Understanding how microorganisms respond to acid pH Is central to their control and successful exploitation. *Front. Microbiol.* **2020**, *11*, 556140. [CrossRef]
36. Jin, Q.; Kirk, M.F. pH as a primary control in environmental microbiology: 1. thermodynamic perspective. *Front. Environ. Sci.* **2018**, *6*, 21. [CrossRef]

microorganisms

MDPI

Article

Detection of Foodborne Pathogens in Acute Gastroenteritis Patient's Stool Samples Using the BioFire® FilmArray® Gastrointestinal PCR Panel in the Republic of Trinidad and Tobago, West Indies

Carelene Lakhan [1,†], Neela Badrie [1,*,†], Adash Ramsubhag [2,†] and Lisa Indar [3,†]

1 Department of Food Production, University of the West Indies, St. Augustine, Trinidad and Tobago
2 Department of Life Sciences, University of the West Indies, St. Augustine, Trinidad and Tobago
3 The Caribbean Public Health Agency, Port of Spain, Trinidad and Tobago
* Correspondence: neela.badrie@sta.uwi.edu; Tel.: +1868-662-2002 (ext. 83211); Fax: +1868-663-9684
† These authors contributed equally to this work.

Abstract: In 2009, the burden of illness study for acute gastroenteritis in Trinidad and Tobago highlighted that ~10% of stool samples tested were positive for a foodborne pathogen. The study also noted that limited laboratory screening for pathogens contributed to a lack of etiology as public health hospitals only routinely tested for *Salmonella* and *Shigella*, and sometimes for *Escherichia coli* and *Campylobacter*. To better understand the foodborne pathogens responsible for acute gastroenteritis, enhanced testing using the BioFire® FilmArray® Gastrointestinal PCR panel was used to screen diarrheal stool samples for 22 pathogens from patients in 2018. The five general public health hospitals (San Fernando, Mt. Hope, Port of Spain, Sangre Grande, and Tobago) were notified of research activities and diarrheal stool samples were collected from all acute gastroenteritis patients. A total of 66 stools were screened and ~30% of samples tested positive for a foodborne pathogen. The current study showed that a much wider range of enteric pathogens were associated with acute gastroenteritis in Trinidad and Tobago than previously reported in 2009. These findings can be used by health officials to guide appropriate interventions, as well as to provide evidence for adoption of the PCR panel detection method at public health hospitals to benefit patient care.

Keywords: diarrhea; molecular diagnostic tool; enhanced foodborne pathogen testing; public health; Trinidad and Tobago

Citation: Lakhan, C.; Badrie, N.; Ramsubhag, A.; Indar, L. Detection of Foodborne Pathogens in Acute Gastroenteritis Patient's Stool Samples Using the BioFire® FilmArray® Gastrointestinal PCR Panel in the Republic of Trinidad and Tobago, West Indies. *Microorganisms* **2022**, *10*, 1601. https://doi.org/10.3390/microorganisms10081601

Academic Editor: Abiodun Adewale Adesiyun

Received: 25 July 2022
Accepted: 29 July 2022
Published: 9 August 2022

1. Introduction

Acute gastroenteritis is an inflammation of the gastrointestinal tract that may result in the sudden onset of symptoms such as abdominal pain, cramping, nausea, vomiting, and diarrhea. This illness, ranging from mild to severe, is usually caused by consumption of food contaminated with harmful bacteria, viruses, parasites, or chemical substances and is responsible for significant cases of morbidity, disability, and mortality globally [1,2]. The World Health Organization (WHO) estimated that approximately one in ten persons globally fall ill and more than 6 billion global cases of diarrheal illness occur annually [3]. The illness was also reported to affect nearly 135,000 residents in Trinidad and Tobago each year and is one of the most flagged syndromes for the Caribbean, causing significant health and economic burdens in the region [4,5] Globally, the most common reported foodborne pathogens transmitted to humans through contaminated food were *Salmonella, Campylobacter, Shigella,* pathogenic *Staphylococcus aureus, Escherichia coli* 0157:H7, *Yersinia, Bacillus, Listeria monocytogenes, Vibrio*, rotavirus, and norovirus [6–8]. Surveillance data from the Caribbean have reported an increase in the number of acute gastroenteritis cases, foodborne disease outbreaks, and related pathogens over the last two decades, with non- typhoidal *Salmonella, Ciguatera* poisoning, *Salmonella typhi, Shigella, Campylobacter,* norovirus, and pathogenic *Escherichia coli* causing most infections [6]. The capacity to accurately identify these causative foodborne pathogens and others is critical to finding or prescribing effective patient treatment, as well as to assist in determining etiology during foodborne outbreaks for quick control and prevention of future occurrences [9].

In 2009, the burden of illness for acute gastroenteritis was estimated in Trinidad and Tobago. In that study, the five general public health hospitals (San Fernando, Mt. Hope, Port of Spain, Sangre Grande, and Tobago) which provide health care to the majority of the population, processed and tested >80% of all acute gastroenteritis ill patient's stool samples; laboratory findings indicated that 10% of all samples tested positive for a foodborne pathogen, and the main method of screening stools was by use of traditional culture practices which provided limited data on the etiology of illness as they only "routinely" tested for *Salmonella* and *Shigella* and "sometimes" tested for *E. coli* and *Campylobacter* based on the physician's request [5]. The use of molecular techniques and other nucleic acid-based methods, even though more costly than traditional culture techniques, were reported to offer highly sensitive and specific automated reliable results with shorter turnaround times [10]. Further analyses involving the use multiplex molecular assays for the detection of common foodborne pathogens in stool samples observed that these panels detected the most gastrointestinal pathogens when compared to conventional methods, and that the Food and Drug Association-approved PCR-based test panel, the BioFire® FilmArray® Gastrointestinal (GI) Panel (BioFire Diagnostics, Salt Lake City, UT, USA), was a top performer among other molecular-based detection tools [10–12]. The BioFire FilmArray test panel was also reported to rapidly detect and identify foodborne pathogen genes in ill patients, which resulted in shorter turn-around times, with accurate, highly sensitive (98.5%), and specific (99.2%) results to benefit patient care [13–15].

A review of current research indicated that there were no follow-up studies conducted after 2009 with regards to enhanced testing for a wider range of foodborne pathogens from acute gastroenteritis patients in Trinidad and Tobago. The ability to test for such pathogens is critical to expand on the etiology of this illness since the current testing method only provides limited data for approximately four (4) pathogens. Use of the BioFire FilmArray PCR panel in this study can be useful to provide the critical etiological data being sought as this PCR panel can test for thirteen (13) foodborne bacteria, five (5) viruses, and four (4) parasites per acute gastroenteritis patient's stool sample. Additionally, literature searches on the use of this PCR panel in Trinidad and Tobago or even the wider Caribbean Region could not be found, and therefore, this research could be considered novel. Hence, the data obtained in this study could be used to assist in guiding health officials to prevent and treat this illness, as well as to provide baseline data for future investigative studies on acute gastroenteritis in Trinidad and Tobago. The objectives of this study were, therefore, to determine by detection the pathogens responsible for acute gastroenteritis through enhanced testing using the PCR panel, and to even further recommend whether its use could be adopted for future testing at public health hospitals in Trinidad and Tobago.

2. Materials and Methods

2.1. Research Design

Diarrheal stool samples were prospectively collected from patients (<5 years old and >5 years old) diagnosed with acute gastroenteritis at the five general public hospitals in Trinidad and Tobago and screened for 22 foodborne pathogens in 2018 using the BioFire FilmArray PCR panel.

2.2. Assumption

This study assumed that the doctor's diagnosis was accurate when patients presented at public health hospitals for acute gastroenteritis based on the case definition of ">3 loose stools in a 24 h period and other typical symptoms of acute gastroenteritis". Therefore, all stool samples submitted and tested were due to this illness.

2.3. Ethical Approval

Ethical approval for this study was granted from the Ethical Review Board of the University of the West Indies and the Review Committee of the Ministry of Health, Trinidad and Tobago. Each participant was informed of the purpose of the study and all data collected were kept confidential.

2.4. Collection of Stool Samples for Pathogen Testing and Sample Size

In January 2018, doctors and nurses at the five general public health hospitals (San Fernando, Mt. Hope, Port of Spain, Sangre Grande, and Tobago) were requested to collect a stool sample from all patients (<5 years old and >5 years old) diagnosed with acute gastroenteritis at their facility for a one-year prospective study. These major hospitals provided a good representative cross-section of the 1,300,000 residents in Trinidad and Tobago who opt to seek healthcare for their illnesses. The case definition for an acute gastroenteritis patient was "the sudden onset of diarrhea, with or without fever (>38 °C or 100.4 °F) and presenting with 3 or more loose/watery stools in the past 24 h, with or without dehydration, vomiting, and/or visible blood". Therefore, in 2018 the sample size included all patients presenting at the public health institution with 3 or more loose/watery stools in a 24 h period. Once the patient agreed to partake in the study, their demographic data along with a fresh stool sample (approximately 1–25 g of liquid stools and/or rectal swabs) were collected and immediately screened for foodborne pathogens according to established laboratory protocols and the manufacturer's recommended methodology for the PCR panel [16]. During the months of January to December 2018, sample collection was sparse, averaging 5–6 samples per month as many persons tend to self-treat for acute gastroenteritis and even with several sensitization sessions held with the public health hospitals' personnel, patients seeking healthcare were still reluctant to submit stool samples, and for those patients who opted to seek private healthcare, they would not have been captured by the public hospitals [17]. A total of 66 diarrheal stools were prospectively collected from patients (<5 years old and >5 years old) diagnosed with acute gastroenteritis at the five general public hospitals and screened for 22 foodborne pathogens in 2018.

2.5. Polymerase Chain Reaction Panel

2.5.1. Foodborne Pathogens for Testing

The pathogens for testing with the BioFire Gastrointestinal Panel were preselected and included: *Campylobacter* (*jejuni*, *coli*, and *upsaliensis*); *Clostridium difficile* (toxin A/B); *Plesiomonas shigelloides*; *Salmonella*; *Yersinia enterocolitica*; *Vibrio* (*parahaemolyticus*, *vulnificus*, and *cholerae*); *Vibrio cholerae*; Enteroaggregative *E. coli* (EAEC); Enteropathogenic *E. coli* (EPEC); Enterotoxigenic *E. coli* (ETEC) lt/st; Shiga-like toxin-producing *E. coli* (STEC) stx1/stx2; *E. coli* O157; Shigella/Enteroinvasive *E. coli* (EIEC); adenovirus F40/41; astrovirus; norovirus GI/GII; rotavirus A; sapovirus (I, II, IV, and V); *Cryptosporidium*; *Cyclospora cayetanensis*; *Entamoeba histolytica*; and *Giardia lamblia*.

2.5.2. Detection of Foodborne Pathogens

The PCR panel is an automated in-vitro diagnostic (IVD) system which uses nested multiplex PCR and high-resolution melting analysis to detect and identify multiple nucleic acid targets from the diarrheal stool samples. Nest multiplex PCR utilizes two stages of PCR. In the first stage, multiple outer primers perform multiplex PCR on the target template that are present in the sample, while the second stage PCR is performed in a singleplex manner, further amplifying the DNA procured during the first stage PCR. The inner primers which are used in the second-stage PCR are made of sequences "nested" within the first-stage PCR products. LC Green® Plus is used as the binding dye which is incorporated into copies of DNA as they are produced during each PCR cycle. After the last stage PCR, the instrument gradually increases the temperature of the reaction to 72 °C and the copies of the double stranded DNA melt. The LC Green® Plus dye is then released and a decrease in the fluorescence is detected by the FilmArray instrument. The time taken to complete the test is <2 h [16].

2.6. Data Analysis

Descriptive statistics to summarize the characteristics of the dataset in terms of foodborne pathogens detected from acute gastroenteritis patients who sought health care were performed using SPSS 24 (IBM Corp., Armonk, NY, USA).

3. Results and Discussion

3.1. Foodborne Pathogens Detected Using the PCR Panel from Stool Samples

In 2018, based on a doctor's accurate diagnosis for acute gastroenteritis, all stool samples from all patients (n = 66) diagnosed with the illness were each tested for 22 pathogens, and nearly 30% (n = 20) of patients tested positive for a foodborne pathogen. The major pathogens detected were *Enteropathogenic E. coli* (~20%), *Campylobacter* (~8%), *Salmonella* (~5%), and *Enteroaggregative E. coli* (~5%), as seen in Figure 1, indicating that there are other pathogens detected to cause acute gastroenteritis illness apart from the ones routinely tested for at the public hospitals. This was the first recorded study in Trinidad and Tobago, and the wider Caribbean Region, on the use of the BioFire Gastrointestinal Panel to test for foodborne pathogens from stool samples submitted by patients ill with acute gastroenteritis.

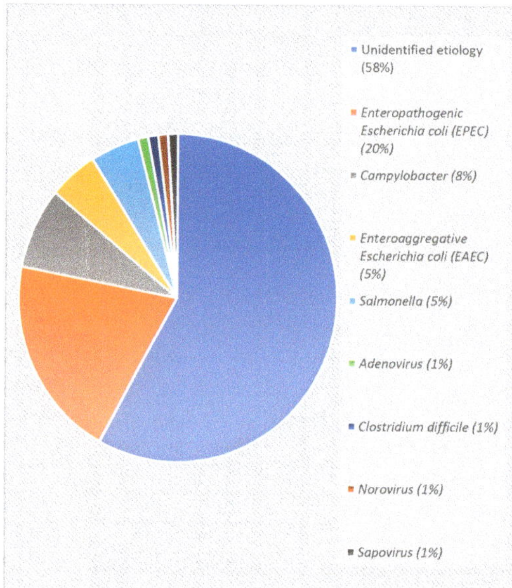

Figure 1. Foodborne pathogens detected using PCR in Trinidad and Tobago, 2018.

3.2. Foodborne Pathogens Responsible for Acute Gastroenteritis Illness in Trinidad and Tobago

Data generated in this research differed in both sample size and detection method from the 2009 study, and while direct comparisons could not be drawn, the pathogens detected from both studies suggested that there are more pathogens responsible for acute gastroenteritis in Trinidad and Tobago than previously reported and currently tested. According to data sourced from the Caribbean Public Health Agency's Laboratory Information System (LABIS), foodborne pathogens reported by the Ministry of Health, Trinidad and Tobago, for the period 2006–2016 included adenovirus, *Campylobacter, E. coli*, norovirus, *Salmonella*, and *Shigella*. Thus, even with a small sample size in 2018, the additional etiological information gathered from this research can still provide a baseline for further studies to assist health officials in guiding appropriate interventions.

Foodborne bacteria were the most prevalent pathogens detected in stool samples from ill patients in this study and were also reported to be responsible for 10% to 55% of all diarrheal cases in developing countries that caused more severe cases of acute gastroenteritis [18,19]. In 2009, *Salmonella* was the leading pathogen responsible for acute gastroenteritis in Trinidad and Tobago, while in 2018, many patients were co-infected with two bacteria spp., namely *EPEC* and *Campylobacter.* EPEC and other mono infections are usually self-limiting but co-infections with other pathogens were reported to increase the severity of this illness [12,20–22].

Identification of foodborne pathogens is crucial to controlling and preventing related illnesses and associated burdens [23]. Data reported in this research and by LABIS indicated that in Trinidad and Tobago, the foodborne pathogens detected to cause acute gastroenteritis were *Salmonella, Shigella,* norovirus, rotavirus, *Campylobacter, Clostridium difficile, E. coli,* adenovirus, and sapovirus. The Centre for Disease Control and Prevention (CDC) reported that, in the USA, adult bacterial related infections were often associated with travel and pathogens such as *Salmonella, Campylobacter, Shigella,* and Shiga toxin-producing *Escherichia coli* (enterohemorrhagic *E. coli)* were identified as the cause [24]. Norovirus was identified as the leading cause of viral gastroenteritis in the adult population, while in children most cases (70%) were caused by rotavirus and norovirus; ~10% by protozoa; and ~10% to 20% by bacteria [25–27]. One diarrheal disease study in children also reported that even though bacterial pathogens were the major group of organisms responsible for the disease, rotavirus was still the single largest causative agent [28]. Generally, many foodborne pathogens can be transmitted in the same way, including animal to human interaction, consuming contaminated food or drink, ill food handlers, and transfer by contaminated surfaces and utensils [29,30]. In addition, with increased international travel and globalization of the food industry, all foodborne pathogens should be monitored and appropriate preventative strategies such as adequate sanitation, food hygiene, targeted health education, and access to safe water for drinking should be encouraged [31,32].

Our research also found that most (58%) stool samples tested did not have a definitive causal agent. This finding was consistent with reports by the CDC in that unidentified etiology was reported to contribute to more episodes of acute gastroenteritis and foodborne illnesses than detected pathogens [33]. Possible reasons for unidentified causal agents in Trinidad and Tobago may include limited pathogen screening as almost 31 foodborne enteric pathogens were identified and reported to cause illness and our study screened for 22 [34]. Hence, pathogens such as *Bacillus cereus, Brucella* spp., *Listeria monocytogenes, Mycobacterium bovis, Staphylococcus aureus, Streptococcus* spp. group A, *Toxoplasma gondii, Trichinella* spp., and hepatitis A virus, which were not tested for in our study, could have been other causative agents. Other possibilities for not identifying a causal agent include that: it may not be a pathogen in the first place but perhaps a toxin or some underlying condition of the patient; the casual organism was no longer present within the patient's system since diarrhea is the body's mechanism to expel toxic materials; or the agent responsible has not yet been identified or remains unproven as causing a foodborne illness [35–38]. Therefore, with continuous surveillance, research activities, and testing for a wider range of pathogens, the gap between identified and unidentified could be narrowed.

3.3. Use of the PCR Panel in Public Health Hospitals, Trinidad and Tobago

The detection rate for foodborne pathogens in our study was lower than that reported in other studies, but so was our sample size [11,39–42]. Nevertheless, this PCR panel detected a positive pathogen in ~30% of all samples tested in less than 2 h. The detection of sapovirus in this study was also noteworthy as limited to no data currently exist for this pathogen as an important cause of acute gastroenteritis in Trinidad and Tobago and considering the growing awareness for sapoviruses in child healthcare, further investigative studies are needed [43]. The traditional culture testing method presently employed at the public health hospitals generally yields results after 24 h and was reported in past studies to be erroneous, time consuming, and at the discretion of physicians' request [44–46]. Use of this PCR panel was also described to rapidly and accurately identify causal pathogens and coinfections, which resulted in the administration of treatments sooner; reduced antibiotic use for "just in case" scenarios; targeted antibiotics therapy started sooner; shorter hospital stays; and reduced morbidity/mortality associated with foodborne illnesses [13,15,47]. Further investigations regarding turnaround times for collection of stools from patients and initiating therapy with cost analysis are also needed, but until then, this PCR panel can be introduced and used at all public health hospitals in Trinidad and Tobago to benefit patient care.

3.4. Demographic Factors and Foodborne Pathogens Detected from Acute Gastroenteritis Patients in Trinidad and Tobago

In 2018, more samples were submitted from acute gastroenteritis patients who were >5 years old, males, and attended the Sangre Grande General Hospital (Table 1). Eight of the positive patients were co-infected with two pathogens, and most co-infections were observed from >5-year-old males who sought health care at the Sangre Grande General Hospital (Table 2). In our study, 100% of co-infections observed involved EPEC and another bacterium. This finding was higher than other studies involving two or more organisms detected which involved EPEC and EAEC [15,41,48]. Similar to our study, one study involving ten participating clinical microbiology laboratories in Austria, Finland, France, Germany, Greece, Ireland, Italy, Portugal, Romania, and the UK, and other studies in Costa Rica and South India, also commonly observed high co-infection rates of *Campylobacter* with EPEC [49–51]. However, such studies failed to provide explanations for these findings or whether microbes were more likely or less likely to occur together or if cases were population specific [20,52–55]. Thus, high coinfection rates indicated varied but common etiology for acute gastroenteritis in Trinidad and Tobago, which warrants further investigations using larger datasets to explain these findings. All *Campylobacter* and ~62% of *EPEC* pathogens detected were from patients who sought health care at the Sangre Grande General Hospital (Table 2). The catchment population for the Sangre Grande General Hospital is ~120,000 residents spanning from Matelot in the North to Guayaguayare, Rio Claro, and Brothers Road in the South to Valencia in the East and covering approximately one-third of the island. This area is sparsely populated and some districts within are considered to have poverty levels higher than the national level. Two-thirds of this region are rural with agriculture (fishing and farming, including chicken farms) as major activities [56,57]. In 2018, the National Meteorological Station reported major flooding activities that affected many residents in this area [58]. All these factors may increase the risk of these residents to *Campylobacter* infections since multiple studies have reported more human *Campylobacter* infections in rural areas when compared to urban, with occupation (agricultural related work); handling of livestock (chickens, cattle); and exposure to contaminated water sources from flooding episodes as source attributions [59–63]. Strategies suggested by the Center of Disease Control and Prevention (CDC) to mitigate the risks of infection in a catchment area like this include targeted educational campaigns such as disease awareness and proper hygiene techniques among agricultural workers; cleaning and sanitizing of hands, tools, and equipment; and not drinking untreated water, especially after flooding episodes. However, further investigations with a larger sample size are recommended to determine whether associations between locations and pathogens occur as samples were collected during different months in 2018 and could suggest infection was not a cluster with one source.

Table 1. Demographic factors associated with foodborne pathogens detected via PCR panel from acute gastroenteritis patients in Trinidad and Tobago, 2018.

Demographics Factors		Foodborne Pathogens Detected via PCR in Trinidad and Tobago, 2018 (n = 28)							
		Campylobacter	*Clostridium difficile*	*Salmonella*	*EAEC*	*EPEC*	*Adenovirus*	*Norovirus*	*Sapovirus*
Age group	<5	1	0	0	2	4	0	0	0
	>5	4	1	3	1	9	1	1	1
Sex	Female	1	0	2	0	5	0	0	1
	Male	4	1	1	3	8	1	1	0
Public Health Hospital	Sangre Grande	5	0	1	2	8	0	0	0
	Mt. Hope	0	1	1	0	2	1	1	1
	Port of Spain	0	0	0	1	3	0	0	0
	San Fernando	0	0	1	0	0	0	0	0
	Tobago $^{\Phi}$	-	-	-	-	-	-	-	-

$^{\Phi}$ No samples received from the Tobago General Hospital.

Table 2. Co-infections detected via PCR panel from acute gastroenteritis patients in Trinidad and Tobago, 2018.

Demographic Factors			Co-Infections Detected via PCR from Acute Gastroenteritis in Trinidad and Tobago, 2018 (n = 8)
Age Group	**Sex**	**Public Health Hospitals**	**Foodborne Pathogens**
>5 years	Male	Sangre Grande	EPEC and *Campylobacter*
<5 years	Female	Sangre Grande	EPEC and *Campylobacter*
>5 years	Male	Sangre Grande	EPEC and *Campylobacter*
>5 years	Male	Sangre Grande	EPEC and *Campylobacter*
>5 years	Male	Sangre Grande	EPEC and *Campylobacter*
>5 years	Female	Sangre Grande	EPEC and *Salmonella*
<5 years	Male	Mt. Hope	EPEC and EAEC
<5 years	Male	Sangre Grande	EPEC and EAEC

Enteropathogenic *Escherichia coli* (EPEC) is a gram-negative bacterium that adheres to intestinal epithelial cells and causes diarrhea, with outbreaks reported mainly among children in pediatric wards and day-care centers [64–67]. Even though typical strains of EPEC were reported to cause infantile diarrhea within the community and hospital settings, in our study nearly 70% of patients were over the age of 5 years, which may suggest new possible strains of this pathogen-with both humans and animals as reservoirs, and therefore, further investigations with larger sample sizes are needed to confirm if this is the situation in Trinidad and Tobago [68–71]. Additionally, EPEC strains in Trinidad and Tobago should be further characterized to determine whether new pathotype or atypical EPEC strains exist. Targeted efforts to reduce infection rates by this pathogen in Trinidad and Tobago include proper hygiene, especially when interacting with animals.

4. Conclusions

Acute gastroenteritis can affect any person at any time along the farm to table continuum but may be prevented once proper food safety systems are in place. To better understand and guide appropriate interventions to prevent this illness, routine monitoring; surveillance activities; and identification of pathogens are essential. In our study, even though the sample size was small, there were other pathogens detected to cause acute gastroenteritis illness apart from the ones routinely tested for at the public hospitals. The pathogens identified in this research that caused illness included: *Salmonella, Shigella,* norovirus, rotavirus, *Campylobacter, Clostridium difficile, E. coli*, adenovirus, and sapovirus. It was observed that more samples were submitted from acute gastroenteritis patients who were >5 years old, males, and attended the Sangre Grande General Hospital. Further investigative studies are required to establish any relationship between location and pathogens detection from ill patients seeking healthcare.

All foodborne pathogens detected can be transmitted in similar ways, including by animal to human interaction, consuming contaminated food or drink, ill food handlers and transfer by contaminated surfaces and utensils. To mitigate the risk of this illness, appropriate preventative strategies such as adequate sanitation, food hygiene, targeted health education, and access to safe water for drinking are needed. This research was the first recorded study in Trinidad and Tobago and the wider Caribbean Region on the use of the BioFire gastrointestinal panel to test for foodborne pathogens from stool samples submitted by patients ill with acute gastroenteritis. This PCR panel allowed for the fast detection of a wide range of pathogens including sapovirus, for which there are no reported studies conducted in Trinidad and Tobago examining it as a cause of acute gastroenteritis, as well as coinfections with two pathogens which would probably not have been detected given the present diagnostic method at the public health hospitals. Hence, data from this study can be used as a baseline in future acute gastroenteritis etiological studies, provide evidence for health officials to guide interventions for the prevention of future foodborne related illnesses, and recommend the adoption and use of this PCR panel at each public hospital to allow for better patient care in terms of the administration of appropriate treatments sooner and ultimately an overall decrease in morbidity/mortality associated with foodborne illnesses in Trinidad and Tobago.

5. Limitations

Due to ethical reasons and patient confidentiality, patient data/history were limited and as such more rigorous analyses/inferences could not be made.

Author Contributions: Conceptualization, N.B., L.I., A.R. and C.L.; methodology, N.B., L.I., A.R. and C.L.; software, N.B., L.I., A.R. and C.L.; validation, N.B., L.I., A.R. and C.L.; formal analysis, N.B., L.I., A.R. and C.L.; investigation, N.B., L.I., A.R. and C.L.; resources, N.B., L.I., A.R. and C.L.; data curation, N.B., L.I., A.R. and C.L.; writing—original draft preparation, N.B., L.I., A.R. and C.L.; writing—review and editing, N.B., L.I., A.R. and C.L.; visualization, N.B., L.I., A.R. and C.L.; supervision, N.B., L.I., A.R. and C.L.; project administration, N.B., L.I., A.R. and C.L.; funding acquisition, N.B., L.I., A.R. and C.L. All authors have read and agreed to the published version of the manuscript.

Funding: This research was funded by the University of the West Indies, St. Augustine Campus Research and Publication Fund Committee—Research Grant CRP #5. Financial contributions were also made by the Caribbean Eco-health Programme (CEHP), Teasdale-Corti Global Health Research Partnership Programme, and International Development Research Center (IRDC).

Institutional Review Board Statement: The study was conducted in accordance with the Declaration of Helsinki and approved by the Ethics Committees of the Ministry of Health, Trinidad and Tobago (approved in 2008), and the University of the West Indies, St. Augustine, Trinidad (approved in 2009), for studies involving humans.

Informed Consent Statement: Informed consent was obtained from all subjects involved in the study.

Data Availability Statement: The data presented in this study are available upon request from the corresponding author. The data are not publicly available due to ethical reasons involving patient confidentiality.

Acknowledgments: All authors would like to thank the staff of: the Ministry of Health in the Republic of Trinidad and Tobago; the Caribbean Public Health Agency Port of Spain; and the University of the West Indies for materials used in experiments and for all the administrative and technical support.

Conflicts of Interest: The authors declare no conflict of interest. The funders had no role in the design of the study; in the collection, analyses, or interpretation of data; in the writing of the manuscript; or in the decision to publish the results.

Abbreviations

The following abbreviations are used in this manuscript:

PCR	Polymerase chain reaction
PCR panel	BioFire® FilmArray® Gastrointestinal Panel Multiplex PCR Test

References

1. Global Burden of Disease (GBD) 2016 Causes of Death Collaborators. Global, regional, and national age-sex specific mortality for 264 causes of death, 1980-2016: A systematic analysis for the Global Burden of Disease Study 2016. *Lancet* **2017**, *390*, 1151–1210. [CrossRef]
2. Papadopoulos, T.; Klamer, S.; Jacquinet, S.; Catry, B.; Litzroth, A.; Mortgat, L.; Mamouris, P.; Rebolledo, J.; Vaes, B.; Van Cauteren, D.; et al. The health and economic impact of acute gastroenteritis in Belgium, 2010–2014. *Epidemiol. Infect.* **2019**, *147*, e146. [CrossRef] [PubMed]
3. World Health Organization (WHO). Newsroom-Fact-Sheets: Food Safety Key Facts. 2021. Available online: https://www.who.int/NEWS-ROOM/FACT-SHEETS/DETAIL/FOOD-SAFETY (accessed on 9 February 2022).
4. Caribbean Public Health Agency (CARPHA). Monthly Syndromic Surveillance Summary. 2017. Available online: https://carpha.org/Portals/0/Documents/MSSS_No3_EpiWeek_8-11-2017.pdf (accessed on 1 June 2021).
5. Lakhan, C.; Badrie, N.; Ramsubhag, A.; Sundaraneedi, K.; Indar, L. Burden and impact of acute gastroenteritis and foodborne pathogens in Trinidad and Tobago. *J. Health Popul. Nutr.* **2013**, *31*, S30.
6. Caribbean Public Health Agency (CARPHA). *State of Public Health in the Caribbean Region 2014–2016: Building Resilience to Immediate and Increasing Threats: Vector-Borne Diseases and Childhood Obesity*; Caribbean Public Health Agency: Port of Spain, Trinidad & Tobago, 2017; ISBN 978-976-8114-38-9.
7. Bhunia, A.K. *Foodborne Microbial Pathogens: Mechanisms and Pathogenesis*; Springer: Berlin/Heidelberg, Germany, 2018.
8. Pires, S.M.; Desta, B.N.; Mughini-Gras, L.; Mmbaga, B.T.; Fayemi, O.E.; Salvador, E.M.; Gobena, T.; Majowicz, S.E.; Hald, T.; Hoejskov, P.S.; et al. Burden of foodborne diseases: Think global, act local. *Curr. Opin. Food Sci.* **2021**, *39*, 152–159. [CrossRef] [PubMed]
9. Centre for Disease Control and Prevention (CDC). Foodborne Outbreaks: You Can Help Solve Foodborne Outbreaks. 2022. Available online: https://www.cdc.gov/foodsafety/outbreaks/investigating-outbreaks/help-solve-outbreaks.html (accessed on 24 April 2022).
10. Law, J.W.; Ab Mutalib, N.S.; Chan, K.G.; Lee, L.H. Rapid methods for the detection of foodborne bacterial pathogens: Principles, applications, advantages and limitations. *Front. Microbiol.* **2015**, *5*, 770. [CrossRef] [PubMed]
11. Huang, R.S.; Johnson, C.L.; Pritchard, L.; Hepler, R.; Ton, T.T.; Dunn, J.J. Performance of the Verigene®enteric pathogens test, Biofire FilmArray™ gastrointestinal panel and Luminex xTAG® gastrointestinal pathogen panel for detection of common enteric pathogens. *Diagn. Microbiol. Infect. Dis.* **2016**, *86*, 336–339. [CrossRef] [PubMed]
12. Zhang, S.X.; Zhou, Y.M.; Xu, W.; Tian, L.G.; Chen, J.X.; Chen, S.H.; Dang, Z.S.; Gu, W.P.; Yin, J.W.; Serrano, E.; et al. Impact of co-infections with enteric pathogens on children suffering from acute diarrhea in southwest China. *Infect. Dis. Poverty* **2016**, *5*, 64. [CrossRef]
13. Foddai, A.; Grant, I.R. Methods for detection of viable foodborne pathogens: Current state-of-art and future prospects. *Appl. Microbiol. Biotechnol.* **2020**, *104*, 4281–4288. [CrossRef]
14. Maurer, F.P.; Christner, M.; Hentschke, M.; Rohde, H. Advances in Rapid Identification and Susceptibility Testing of Bacteria in the Clinical Microbiology Laboratory: Implications for Patient Care and Antimicrobial Stewardship Programs. *Infect. Dis. Rep.* **2017**, *9*, 6839. [CrossRef] [PubMed]
15. Torres-Miranda, D.; Akselrod, H.; Karsner, R.; Secco, A.; Silva-Cantillo, D.; Siegel, M.O.; Roberts, A.D.; Simon, G.L. Use of BioFire FilmArray gastrointestinal PCR panel associated with reductions in antibiotic use, time to optimal antibiotics, and length of stay. *BMC Gastroenterol.* **2020**, *20*, 246. [CrossRef]
16. FilmArray®Panels, Gastrointestinal Panel. Documents: Gastrointestinal Tech Notes. 2018. Available online: https://www.biofiredx.com/products/the-filmarray-panels/#gastrointestinal (accessed on 11 January 2021).
17. Rudzik, A.E. Examining health equity through satisfaction and confidence of patients in primary healthcare in the Republic of Trinidad and Tobago. *J. Health Popul. Nutr.* **2003**, *21*, 243–250. [PubMed]
18. Bintsis, T. Foodborne pathogens. *AIMS Microbiol.* **2017**, *3*, 529–563. [CrossRef] [PubMed]
19. Sattar, S.B.A.; Singh, S. *Bacterial Gastroenteritis*; StatPearls Publishing: Treasure Island, FL, USA, 2021. Available online: https://www.ncbi.nlm.nih.gov/books/NBK513295/ (accessed on 1 June 2021).
20. Bhavnani, D.; Goldstick, J.E.; Cevallos, W.; Trueba, G.; Eisenberg, J.N. Synergistic effects between rotavirus and coinfecting pathogens on diarrheal disease: Evidence from a community-based study in northwestern Ecuador. *Am. J. Epidemiol.* **2012**, *176*, 387–395. [CrossRef]
21. Valentini, D.; Vittucci, A.C.; Grandin, A.; Tozzi, A.E.; Russo, C.; Onori, M.; Menichella, D.; Bartuli, A.; Villani, A. Coinfection in acute gastroenteritis predicts a more severe clinical course in children. *Eur. J. Clin. Microbiol. Infect. Dis.* **2013**, *32*, 909–915. [CrossRef] [PubMed]

22. Vergadi, E.; Maraki, S.; Dardamani, E.; Ladomenou, F.; Galanakis, E. Polymicrobial Gastroenteritis in Children. *Acta Paediatr.* **2021**, *110*, 2240–2245. [CrossRef]
23. Wei, X.; Zhao, X. Advances in typing and identification of foodborne pathogens. *Curr. Opin. Food Sci.* **2021**, 52–57. [CrossRef]
24. Centre for Disease Control and Prevention (CDC). Foodborne Germs and Illnesses: Causes of Food Poisoning. 2020. Available online: https://www.cdc.gov/foodsafety/foodborne-germs.html (accessed on 1 June 2021).
25. Chow, C.M.; Leung, A.K.C.; Hon, K.L. Acute gastroenteritis: From guidelines to real life. *Clin. Exp. Gastroenterol.* **2010**, *3*, 97–112. [CrossRef] [PubMed]
26. Krenzer, M.E. Viral gastroenteritis in the adult population: The GI peril. *Crit. Care Nurs. Clin. N. Am.* **2012**, *24*, 541–553. [CrossRef]
27. O'Shea, H.; Blacklaws, B.A.; Collins, P.J.; McKillen, J.; Fitzgerald, R. Viruses Associated with Foodborne Infections. *Ref. Modul. Life Sci.* **2019**. [CrossRef]
28. Dadonaite, Bernadeta, Hannah Ritchie and Max Roser. Diarrheal Diseases. 2018. Available online: https://ourworldindata.org/diarrheal-diseases (accessed on 1 June 2021).
29. Oliver, S.P. Foodborne Pathogens and Disease, Special Issue on the National and International PulseNet Network. *Foodborne Pathog. Dis.* **2019**, *16*, 439–440. [CrossRef] [PubMed]
30. Stein, R.A.; Chirilã, M. Routes of Transmission in the Food Chain. *Foodborne Dis.* **2017**, 65–103. [CrossRef]
31. FoodSafety.gov (FSGov). Keep Food Safe: Four (4) Steps to Food Safety. 2020. Available online: https://www.foodsafety.gov/keep-food-safe/4-steps-to-foodsafety (accessed on 12 June 2021).
32. World Health Organization (WHO). WHO Fact Sheets—Diarrhoeal Disease 2017. Available online: https://www.who.int/news-room/fact-sheets/detail/diarrhoeal-disease (accessed on 11 January 2021).
33. Burke, R.M.; Mattison, C.P.; Marsh, Z.; Shioda, K.; Donald, J.; Salas, S.B.; Naleway, A.L.; Biggs, C.; Schmidt, M.A.; Hall, A.J. *Norovirus* and Other Viral Causes of Medically Attended Acute Gastroenteritis Across the Age Spectrum: Results from the Medically Attended Acute Gastroenteritis Study in the United States. *Clin. Infect. Dis.* **2021**, *4*, e913–e920. [CrossRef] [PubMed]
34. Adley, C.C.; Ryan, M.P. The Nature and Extent of Foodborne Disease. In *Antimicrobial Food Packaging*; Academic Press: Cambridge, MA, USA, 2016; pp. 1–10. ISBN 9780128007235. [CrossRef]
35. Franklin, N.; Hope, K.; Glasgow, K.; Glass, K. Describing the epidemiology of foodborne outbreaks in New South Wales from 2000 to 2017. *Foodborne Pathog. Dis.* **2020**, *17*, 701–711. [CrossRef] [PubMed]
36. Jahan, S. Epidemiology of foodborne illness. *Sci. Health Soc. Asp. Food Ind.* **2012**, *1*, 321–342.
37. Scallan, E.; Griffin, P.M.; Angulo, F.J.; Tauxe, R.V.; Hoekstra, R.M. Foodborne Illness Acquired in the United States—Unspecified Agents. *Emerg. Infect. Dis.* **2011**, *17*, 16–22. [CrossRef] [PubMed]
38. Smith, J.L.; Fratamico, P.M. Emerging and re-emerging foodborne pathogens. *Foodborne Pathog. Dis.* **2018**, *15*, 737–757. [CrossRef]
39. Murphy, C.N.; Fowler, R.C.; Iwen, P.C.; Fey, P.D. Evaluation of the BioFire FilmArray gastrointestinal panel in a midwestern academic hospital. *Eur. J. Clin. Microbiol. Infect. Dis.* **2016**, *36*, 747. [CrossRef]
40. Park, S.; Hitchcock, M.M.; Gomez, C.A.; Banaei, N. Is follow-up testing with the FilmArray gastrointestinal multiplex PCR panel necessary? *J. Clin. Microbiol.* **2017**, *55*, 1154–1161. [CrossRef]
41. Piralla, A.; Lunghi, G.; Ardissino, G.; Girello, A.; Premoli, M.; Bava, E.; Arghittu, M.; Colombo, M.R.; Cognetto, A.; Bono, P.; et al. FilmArray™ GI panel performance for the diagnosis of acute gastroenteritis or hemorragic diarrhea. *BMC Microbiol.* **2017**, *17*, 111. [CrossRef] [PubMed]
42. Stockmann, C.; Pavia, A.T.; Graham, B.; Vaughn, M.; Crisp, R.; Poritz, M.A.; Rogatcheva, M. Detection of 23 gastrointestinal pathogens among children who present with diarrhea. *J. Pediatric Infect. Dis. Soc.* **2016**, *6*, 231–238. [CrossRef]
43. Becker-Dreps, S.; Bucardo, F.; Vinjé, J. Sapovirus: An important cause of acute gastroenteritis in children. *Lancet Child Adolesc. Health* **2019**, *3*, 758–759. [CrossRef]
44. Machiels, J.D.; Cremers, A.; van Bergen-Verkuyten, M.; Paardekoper-Strijbosch, S.; Frijns, K.; Wertheim, H.; Rahamat-Langendoen, J.; Melchers, W. Impact of the BioFire FilmArray gastrointestinal panel on patient care and infection control. *PLoS ONE* **2020**, *15*, e0228596. [CrossRef]
45. Priyanka, B.; Patil, R.K.; Dwarakanath, S. A review on detection methods used for foodborne pathogens. *Indian, J. Med. Res.* **2016**, *144*, 327–338. [CrossRef]
46. Valledor, S.; Valledor, I.; Gil-Rodríguez, M.C.; Seral, C.; Castillo, J. Comparison of several Real-Time PCR Kits versus a Culture-dependent Algorithm to Identify Enteropathogens in Stool Samples. *Sci. Rep.* **2020**, *10*, 4301. [CrossRef] [PubMed]
47. Melendez, J.H.; Frankel, Y.M.; An, A.T.; Williams, L.; Price, L.B.; Wang, N.Y.; Lazarus, G.S.; Zenilman, J.M. Real-time PCR assays compared to culture-based approaches for identification of aerobic bacteria in chronic wounds. *Clin. Microbiol. Infect.* **2010**, *16*, 1762–1769. [CrossRef]
48. Freeman, K.; Mistry, H.; Tsertsvadze, A.; Royle, P.; McCarthy, N.; Taylor-Phillips, S.; Manuel, R.; Mason, J. Multiplex tests to identify gastrointestinal bacteria, viruses and parasites in people with suspected infectious gastroenteritis: A systematic review and economic analysis. *Health Technol. Assess.* **2017**, *21*, 1–188. [CrossRef]
49. Murugan, D.; Anandan, S.; Veereraghavan, B. Real-time multiplex PCR assay reveals the increased prevalence of *Campylobacter* spp and diarrhoeagenic *Escherichia coli* in humans from Vellore, South India. *J. Med. Microbiol.* **2022**, 71. [CrossRef]
50. Pérez-Corrales, C. and Leandro-Sandí, K. Diarrheagenic Escherichia coli in Costa Rican children: A 9-year retrospective study. *BMC Res. Notes* **2019**, *12*, 297. [CrossRef]

51. Spina, A.; Kerr, K.G.; Cormican, M.; Barbut, F.; Eigentler, A.; Zerva, L.; Tassios, P.; Popescu, G.A.; Rafila, A.; Eerola, E.; et al. Spectrum of enteropathogens detected by the FilmArray GI Panel in a multicentre study of community-acquired gastroenteritis. *Clin. Microbiol. Infect.* **2015**, *21*, 719–728. [CrossRef]
52. Andersson, M.; Kabayiza, J.C.; Elfving, K.; Nilsson, S.; Msellem, M.I.; Mårtensson, A.; Björkman, A.; Bergström, T.; Lindh, M. Coinfection with Enteric Pathogens in East African Children with Acute Gastroenteritis-Associations and Interpretations. *Am. J. Trop. Med. Hyg.* **2018**, *98*, 1566–1570. [CrossRef] [PubMed]
53. Ardissino, G.; Possenti, I.; Salardi, S.; Tel, F.; Colombo, E.; Testa, S.; Daprai, L.; Picicco, D.; Colombo, R.M.; Torresani, E. Co-infection in children with bloody diarrhea caused by Shiga toxin–producing Escherichia coli: Data of the North Italian HUS Network. *J. Pediatric Gastroenterol. Nutr.* **2014**, *59*, 218–220. [CrossRef] [PubMed]
54. Klein, E.; Stapp, J.; Clausen, C.; Boster, D.; Wells, J.; Qin, X.; Swerdlow, D.; Tarr, P.J. Shiga toxin-producing *Escherichia coli* in children with diarrhea: A prospective point-of-care study. *Pediatrics* **2002**, *141*, 172–178. [CrossRef] [PubMed]
55. Lindsay, B.; Ramamurthy, T.; Gupta, S.S.; Takeda, Y.; Rajendran, K.; Nair, G.B.; Stine, O.C. Diarrheagenic pathogens in polymicrobial infections. *Emerg. Infect. Dis.* **2011**, *17*, 606. [CrossRef]
56. Adewakun, A.A.; Percival, T.M.; Barclay, S.R.; Amaechi, B.T. Caries status of children in eastern Trinidad, West Indies. *Oral. Health Prev. Dent.* **2005**, *3*, 249–261. [PubMed]
57. Esnard-Flavius, T.; Aziz, Z. Microcredit, Microenterprises and Social Welfare of the Rural Poor in North-Eastern Trinidad: An Evaluation Of "Hope". *Asian Acad. Manag. J.* **2011**, *1*, 95–118.
58. Trinidad and Tobago Meteorological Services (TTMS). Devastating Floods of October 2018: The Mother of All Floods. 2018. Available online: http://www.cmo.org.tt/Presentations/2018/Trinidad_and_Tobago-2018.pdf (accessed on 12 June 2021).
59. Kabore, H.; Levallois, P.; Michel, P.; Payment, P.; Dery, P.; Gingras, S. Association between potential zoonotic enteric infections in children and environmental risk factors in Quebec, 1999–2006. *Zoonoses Public Health* **2010**, *57*, e195–e205. [CrossRef]
60. Levesque, S.; Fournier, E.; Carrier, N.; Frost, E.; Arbeit, R.D.; Michaud, S. Campylobacteriosis in urban versus rural areas: A case-case study integrated with molecular typing to validate risk factors and to attribute sources of infection. *PLoS ONE* **2013**, *8*, e83731. [CrossRef] [PubMed]
61. Popovic-Uroic, T. *Campylobacter jejuni* and *Campylobacter coli* diarrhoea in rural and urban populations in Yugoslavia. *Epidemiol. Infect.* **1989**, *102*, 59–67. [CrossRef]
62. Rodríguez-Lázaro, D.; Cook, N.; Ruggeri, F.M. Virus hazards from food, water, and other contaminated environments. *FEMS Microbiol. Rev.* **2012**, *36*, 786–814. [CrossRef]
63. Su, C.P.; de Perio, M.A.; Fagan, K.; Smith, M.L.; Salehi, E.; Levine, S.; Gruszynski, K.; Luckhaupt, S.E. Luckhaupt. Occupational distribution of Campylobacteriosis and Salmonellosis cases—Maryland, Ohio, and Virginia, 2014. *MMWR-Morb. Mortal. Wkly. Rep.* **2017**, *66*, 850. [CrossRef] [PubMed]
64. Croxen, M.A.; Law, R.J.; Scholz, R.; Keeney, K.M.; Wlodarska, M.; Finlay, B.B. Recent advances in understanding enteric pathogenic *Escherichia coli*. *Clin. Microbiol. Rev.* **2013**, *26*, 822–880. [CrossRef]
65. Dedeić-Ljubović, A.; Hukić, M.; Bekić, D.; Zvizdić, A. Frequency and distribution of diarrhoeagenic Escherichia coli strains isolated from pediatric patients with diarrhoea in Bosnia and Herzegovina. *Bosn. J. Basic Med. Sci.* **2009**, *9*, 148–155. [CrossRef]
66. Huiwen Deborah Chen and Gad Frankel. Enteropathogenic Escherichia coli: Unravelling pathogenesis. *FEMS Microbiol. Rev.* **2005**, *29*, 83–98. [CrossRef] [PubMed]
67. Panlozzi, L.J.; Johnson, K.E.; Kamahele, L.M.; Clausen, C.R.; Riley, L.W.; Helgerson, S.D. Diarrhoea associated with adherent enteropathogenic Escherichia coli in an infant and toddler centre, Seattle, Washington. *Pediatrics* **1989**, *77*, 296–300. [CrossRef]
68. Bokhari, H.; Shah, M.A.; Asad, S.; Akhtar, S.; Akram, M.; Wren, B.W. *Escherichia coli* Pathotypes in Pakistan from Consecutive Floods in 2010 and 2011. *Am. Soc. Trop. Med. Hyg.* **2013**, *88*, 519–525. Available online: https://www.ajtmh.org/view/journals/tpmd/88/3/article-p519.xml (accessed on 12 June 2021). [CrossRef]
69. Ochoa, T.J.; Contreras, C.A. Enteropathogenic Escherichia coli infection in children. *Curr. Opin. Infect. Dis.* **2011**, *24*, 478–483. [CrossRef] [PubMed]
70. Shetty, V.A.; Kumar, S.H.; Shetty, A.K.; Karunasagar, I.; Karunasagar, I. Prevalence and characterization of diarrheagenic Escherichia coli isolated from adults and children in Mangalore, India. *J. Lab. Physicians* **2012**, *4*, 24–29. [CrossRef] [PubMed]
71. Trabulsi, L.R.; Keller, R.; Tardelli Gomes, T.A. Typical and atypical enteropathogenic Escherichia coli. *Emerg. Infect. Dis.* **2002**, *8*, 508–513. [CrossRef] [PubMed]

microorganisms

Article

Effect of Biltong Dried Beef Processing on the Reduction of *Listeria monocytogenes*, *E. coli* O157:H7, and *Staphylococcus aureus*, and the Contribution of the Major Marinade Components

Kavya Gavai [1,2], Caitlin Karolenko [1,2] and Peter M. Muriana [1,2,*]

1 Robert M. Kerr Food & Agricultural Products Center, Oklahoma State University, Stillwater, OK 74078, USA; kavya.gavai@okstate.edu (K.G.); caitlin.e.karolenko@okstate.edu (C.K.)
2 Department of Animal and Food Sciences, Oklahoma State University, Stillwater, OK 74078, USA
* Correspondence: peter.muriana@okstate.edu; Tel.: +1-405-744-5563

Abstract: Biltong is a dry beef product that is manufactured without a heat lethality step, raising concerns of whether effective microbial pathogen reduction can occur during biltong processing. Raw beef inoculated with 4-strain cocktails of either *E. coli* O157:H7, *Listeria monocytogenes*, or *Staphylococcus aureus*, and processed with a standard biltong process, were shown to incur a >5-log reduction in 6–8 days after marination by vacuum-tumbling for 30 min in vinegar, salt, spices (coriander, pepper) when dried at 23.9 °C (75 °F) at 55% relative humidity (RH). Pathogenic challenge strains were acid-adapted in media containing 1% glucose to ensure that the process was sufficiently robust to inhibit acid tolerant strains. Internal water activity (A_w) reached < 0.85 at 5-log reduction levels, ensuring that conditions were lower than that which would support bacterial growth, or toxin production by *S. aureus* should it be internalized during vacuum tumbling. This was further confirmed by ELISA testing for staphylococcal enterotoxins A and B (SEA, SEB) after marination and again after 10 days of drying whereby levels were lower than initial post-marination levels. Comparison of log reduction curves obtained for *E. coli* O157:H7, *L. monocytogenes*, *S. aureus*, and *Salmonella* (prior study) showed that microbial reduction was not significantly different ($p < 0.05$) demonstrating that even without a heat lethality step, the biltong process we examined produces a safe beef product according to USDA-FSIS guidelines.

Keywords: *Listeria monocytogenes*; *E. coli* O157:H7; *Staphylococcus aureus*; biltong; log reduction; acid adaptation; water activity

Citation: Gavai, K.; Karolenko, C.; Muriana, P.M. Effect of Biltong Dried Beef Processing on the Reduction of *Listeria monocytogenes*, *E. coli* O157:H7, and *Staphylococcus aureus*, and the Contribution of the Major Marinade Components. *Microorganisms* **2022**, *10*, 1308. https://doi.org/10.3390/microorganisms10071308

Academic Editor: Abiodun Adewale Adesiyun

Received: 7 June 2022
Accepted: 25 June 2022
Published: 28 June 2022

1. Introduction

Dried beef products are found worldwide, and although they may have originated as a way to preserve perishable meat products [1], they have now become accepted as nutritious high protein 'meat snacks' for sportsmen, campers, and hikers [2]. They are also included in specialized 'paleolithic diets' to simulate foods our hunter-gatherer ancestors consumed, serving as a modern means of reducing weight [3]. Popular dried beef products such as beef jerky, kippered beef, and biltong can be found in nearly every supermarket and convenience store. A recent newcomer to the manufacturing of biltong, Stryve Foods (Plano, TX, USA), recently announced record annual sales of $30 million USD (2021), representing an increase of 77% over 2020 and projected sales of $43–48 million USD for 2022. It is just one of many biltong beef processors experiencing increased sales of this new product line in US markets that have been dominated by traditional American style beef jerky.

Although biltong is noted to have originated in South Africa as the result of early Dutch settlers attempting to preserve meat, there are no South African regulatory guidelines for its manufacture. As biltong has grown in popularity in the UK and US, efforts to export the product directly from South Africa have been thwarted. In prior years, it was difficult to export biltong out of South Africa into developed countries where food safety is heavily regulated if there is no acceptable regulatory guidance for its manufacture, unless it was made to EU standards or in a HACCP-approved facility. Kussaga et al. (2014) have suggested that the absence of risk-based food safety management programs such as HACCP or ISO22000 from government through retail/company level has prevented many exports from South Africa at a time when their products are in demand [4]. As is often the case, guidance from the US, New Zealand, or other developed countries on the manufacture of RTE meats such as beef jerky was often cited in making biltong [5]. Similarly, questions posed to British Columbia health authorities (Canada) regarding biltong manufacture relied on information from the Canadian Food Inspection Service (CFIS) which adopted guidance from the United States Department of Agriculture, Food Safety and Inspection Service (USDA-FSIS) [6].

In the US, the manufacture of dried beef products such as beef jerky is regulated by USDA-FSIS and provides industry guidance to ensure product safety. Since biltong is not defined by a standard of identity, USDA-FSIS often cites 'Compliance Guidelines' for conditions of manufacture of beef jerky that includes heating in the presence of 90% RH to govern issues of concern with biltong manufacture [7,8]. Beef jerky is often processed with a high temperature lethality step of >62.8 °C (145 °F) dependent on % RH, whereas biltong is air-dried at room temperature at 21.1–26.7 °C (70–80 °F) after marination with salt, spices, and vinegar [9,10]. Since high temperatures are not used in biltong manufacture, USDA-FSIS has allowed modifications to biltong processing similar to what was allowed for pepperoni and summer sausage at the height of shigatoxigenic *E. coli* outbreaks in the 1990's with such products [11] and currently allows two options to accommodate biltong processing. One option is to test all lots of ingredients prior to use for presence/absence of *Salmonella* spp. as a pathogen of concern, and then use a manufacturing process that demonstrates $\geq$ 2-log reduction of *Salmonella* (i.e., 'a pathogen of concern'). The ingredients should test negative; if they test positive, those ingredients cannot be used unless rendered free of *Salmonella* spp. The second option is to use a process that provides $\geq$5-log reduction of *Salmonella* spp. with no need for *Salmonella* testing of ingredients. This later alternative is preferred because it eliminates costly *Salmonella* testing and problems that arise when *Salmonella*-positive ingredients are encountered. However, the difficulty is in demonstrating that the process achieves $\geq$5-log reduction of the pathogen of concern.

The absence of a heat lethality (cook) step in the processing of biltong has been a concern for pathogens associated with raw beef and/or the beef processing environment. Studies have demonstrated the presence of *Staphylococcus* and *Bacillus* (*Bacillus cereus*) as the predominant taxa associated with biltong obtained from local butcheries in South Africa [12]. Similarly, another study examining 150 samples of biltong purchased in South Africa demonstrated the presence of *Escherichia coli*, coagulase-positive *Staphylococcus*, *Salmonella* and *Listeria monocytogenes* [13]. A lack of information on details of the impact of biltong marinades, vinegar, and drying on pathogen reduction given the moderate processing conditions has caused USDA-FSIS to recognize that there are 'knowledge gaps' with such new products that are being introduced in the US [14]. These knowledge gap areas are often noted as being insufficiently covered by scientific information and they are often targeted as viable research topic areas by USDA or industry-funded research programs.

The USDA-FSIS requires that manufacturers of shelf stable dried beef products validate the microbial safety of their process of manufacture. *Salmonella* spp. has been noted as a 'pathogen of concern' because *Salmonella*, and other pathogens, have historically been associated with food production from animals and the meat derived from them [15–18]. The USDA-FSIS requires process validation with only a single 'pathogen of concern' for commercial processors to manufacture and sell biltong. However, corporate management of retail supermarket chains often require additional food safety assurances that other pathogens also associated with beef (*L. monocytogenes*, *E. coli* O157:H7, and *S. aureus*) are also effectively inhibited by the manufacturing process [13,19,20]. Currently, there is limited data ('data gaps') regarding biltong processes that demonstrate sufficient reduction of these foodborne pathogens.

In this study, we examined a common biltong process for process lethality against acid-adapted cultures of *E. coli* O157:H7, *L. monocytogenes*, and *S. aureus*, the effect of individual ingredient components (salt, spice, vinegar) on process lethality against *E. coli* O157:H7 and *L. monocytogenes*, and whether biltong processing allows for *S. aureus* enterotoxin production from *S. aureus* inoculated beef. This work, together with our previous work with *Salmonella* spp., should satisfy process safety concerns regarding the four major pathogens associated with dried beef products and help fill scientific data gaps that exist for biltong-processed dried meat products.

2. Materials and Methods

2.1. Bacterial Strains, Growth Conditions, and Antibiotic Resistance

Bacterial cultures used in this study included 4 strains each of *L. monocytogenes*, *E. coli* O157:H7, and *S. aureus* (Table 1). Strains of *L. monocytogenes* included ATCC 49594/Scott A-2 (serotype 4b, human isolate), V7-2 (serotype 1/2a, milk isolate), 39-2 (retail hotdog isolate), and 383-2 (ground beef isolate) [21]. These strains were resistant to streptomycin (100 ug/mL; Sigma-Aldrich, St. Louis, MO, USA) and rifamycin S/V (10 µg/mL; Sigma-Aldrich) and were plated on tryptic soy agar (TSA; Difco Brand, Becton-Dickenson, Sparks, MD, USA) containing these antibiotics for enumeration of inoculated cultures recovered from biltong beef. Strains of *E. coli* O157:H7 included ATCC 35150, ATCC 43894, ATCC 43889, and ATCC 45756 that are known for acid tolerance [22,23]. These strains were all resistant to 5 µg/mL novobiocin and 2.5 µg/mL rifamycin S/V (Sigma-Aldrich) and enumeration of these strains was conducted on TSA (Difco) containing these antibiotics. Strains of *S. aureus* and the staphylococcal enterotoxins they produce included ATCC 8095 (SEA), ATCC 13565 (SEA), ATCC 14458 (SEB), and ATCC 51740 (SEB) [24,25]. These strains were resistant to clindamycin (5 µg/mL; Sigma-Aldrich) and were plated on TSA (Difco) containing this antibiotic.

Table 1. List of strains used as challenge organisms for biltong processing in this study.

Organism	Strain Designation	Culture Collection Designation	Source
L. monocytogenes	ATCC 49594	PMM 264	ScottA-2; Clinical isolate
L. monocytogenes	V7-2	PMM 266	Clinical isolate
L. monocytogenes	39-2	PMM 39	Retail hotdogs
L. monocytogenes	383-2	PMM 383	Retail ground beef
E. coli O157:H7	ATCC 35150	PMM 407	Human feces
E. coli O157:H7	ATCC 43889	PMM 1111	Human feces
E. coli O157:H7	ATCC 43894	PMM 405	Human feces
E. coli O157:H7	ATCC 45756	PMM 715	JB Luchansky, USDA-ARS
S. aureus	ATCC 8095	PMM 323	Cream pie
S. aureus	ATCC 13565	PMM 318	Ham, enterotoxin illness
S. aureus	ATCC 14458	PMM 319	Human feces, diarrhea
S. aureus	ATCC 51740	PMM 678	Margarine

The various pathogens were confirmed for typical phenotypes on selective and differential agars: *L. monocytogenes* on modified Oxford agar (MOX, Difco, BD), *E. coli* O157:H7 on CHROMagar O157 (DRG International, Springfield, NJ, USA) and RAPID *E. coli* O157:H7 medium (Bio-Rad Laboratories, Hercules, CA, USA), and *S. aureus* on Mannitol Salt Agar (Difco-BBL, BD Laboratories, Franklin Lakes, NJ, USA); data not shown. The antibiotic resistances of the strains used in this study were also confirmed by plating on TSA, with and without antibiotics, to ensure equivalent enumeration so that survivors from processed biltong could be recovered on antibiotic-containing TSA.

Bacterial cultures were grown in tryptic soy broth (TSB, BD Bacto Brand BD211825, Franklin Lakes, NJ, USA) in 9 mL tubes at 30 °C (*L. monocytogenes*) or 37 °C (*E. coli*, *S. aureus*). Cultures were maintained for storage by centrifugation (6000× *g*, 5 °C) of 9 mL of fresh, overnight cultures and cell pellets were resuspended in 2–3 mL of fresh sterile TSB containing 10% glycerol. Cell suspensions were placed into glass vials and stored in an ultra-low freezer (−80 °C). Frozen stocks were revived by transferring 100 µL of the thawed cell suspension into 9 mL of TSB, incubating overnight at 30 °C or 37 °C, and were sub-cultured twice before use. Microbial enumeration was carried out on tryptic soy agar (TSA, BD Bacto; 1.5% agar) and plated in duplicate.

Bacterial cultures used for inoculation of biltong beef were 'acid adapted' by growing them in media augmented with glucose prior to use according to Wilde et al. [26] and Karolenko et al. [27]. Individual bacterial cultures were first propagated overnight at 37 °C (*E. coli, Staphylococcus*) or 30 °C (*Listeria*) in 9 mL TSB (BD Bacto BD286220). These cultures were individually used to re-inoculate 250 mL centrifuge bottles containing 200 mL pre-warmed TSB containing 1% glucose (BD Bacto BD286220 + 1% glucose) which were again incubated overnight (at 37 °C or 30 °C) for approximately 18 h. Individual cultures in 250 mL bottles were harvested by centrifugation, resuspended with 5 mL of sterile 0.1% buffered peptone water (BPW, BD Difco), mixed in equal proportions, and held refrigerated (5 °C) until use (within 1–2 hr).

2.2. *Beef Handling, Fabrication, and Inoculation*

Boneless beef bottom rounds (outside round, flat; select grade or ungraded) were purchased from a local meat processor (Ralph's, Perkins, OK, USA) who procured them from a broker (Figure 1A). Beef was stored in commercial coolers at the R.M. Kerr Food and Ag Products Center meat pilot plant until needed (i.e., used within 1–2 weeks). Beef was initially trimmed of excess fat, sliced into strips of ~1.9 cm thick, and then cut into small pieces that were ~1.9 cm thick, ~5.1 cm wide, and ~7.6 cm long (~80–100 g) (Figure 1B,C). After bottom rounds were trimmed and cut, the beef was placed on trays, wrapped in plastic bags, and maintained at 4 °C until processed (i.e., next morning).

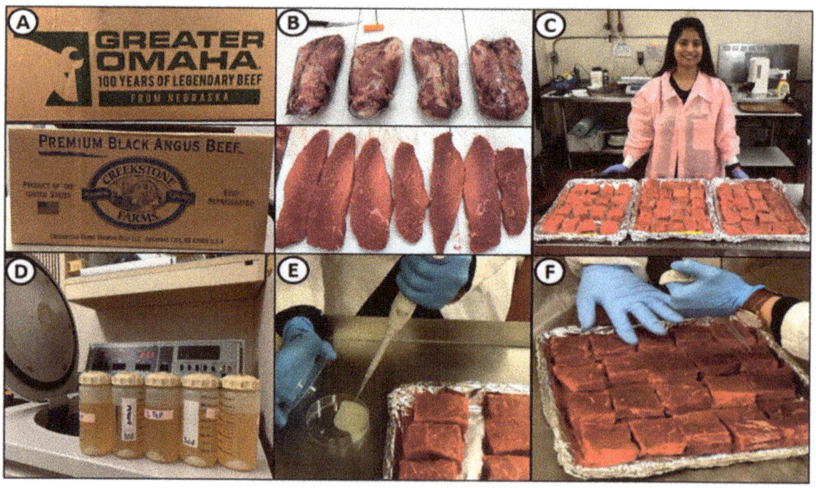

Figure 1. Biltong processing 1: (**A**) boxes of bottom rounds from beef processors; (**B**) bottom rounds in vacuum packaging and sliced after trimming; (**C**) further cutting of sliced biltong beef into small biltong beef pieces; (**D**) recovery and concentration of acid-adapted inoculum strains; (**E**) pipetting inoculum onto beef pieces; (**F**) spreading inoculum on beef surface using a 'gloved finger'.

Inoculation of mixed strains of cultures (i.e., preparation was described earlier; Figure 1D) was performed as follows: (a) the appropriate amount of beef pieces for a particular trial were placed on foil-lined trays and held in the refrigerator until inoculation; (b) ~150 µL of the resuspended/concentrated inoculum mixture was applied by micropipette onto the surface of each beef slice; (c) the inoculum was rubbed over the surface with a double-gloved finger by another person who was assisting; (d) the pieces were then turned over and the process repeated on the other side; (e) the tray(s) of inoculated beef were then placed in a refrigerator 5 °C for at least 30 min to allow attachment (Figure 1E,F).

2.3. Marination of Inoculated Beef Pieces

Inoculated beef pieces that were held at refrigeration for attachment were dipped in sterile water for 30 sec to mimic processes where beef is rinsed or dipped in antimicrobial solution (Figure 2A). A basic marinade mixture was taken as an average of approximately 10 biltong recipes found on the internet and total formulation was comprised of beef (92%), vinegar (4%), salt (2.1%), coriander (1.1%), and black pepper (0.8%) (Figure 2B). Since each non-beef component was based on the beef weight, for each trial we obtained the weight of the total inoculated beef pieces to determine the amount of spices and vinegar to add. These were then mixed with a whisk in a pre-chilled tumbling chamber and the inoculated beef pieces were then added. The tumbler chamber cover was sealed and a vacuum of 15 in Hg was established with a small vacuum pump and the tumbler was set to rotate for 30 min using a Biro VTS-43 tumbler (Biro, Marblehead, OH, USA; Figure 2C,D).

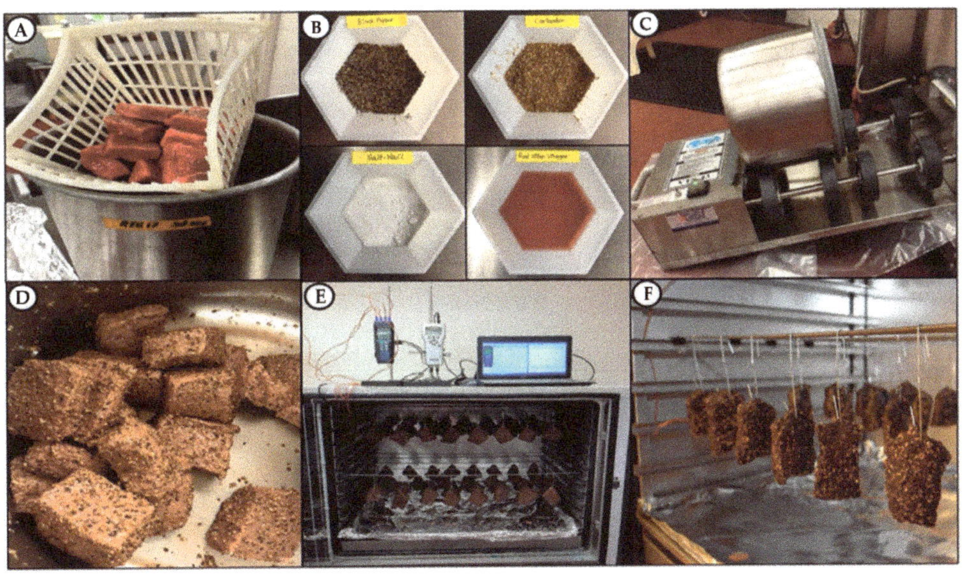

Figure 2. Biltong processing 2: (**A**) dip treatment of inoculated pieces in water (or antimicrobial); (**B**) black pepper, coriander, salt, and vinegar for marinade; (**C**) vacuum tumbling of biltong beef in marinade; (**D**) spiced biltong beef after marination; (**E**) humidity oven with hanging biltong beef and handheld temperature and humidity monitors; (**F**) biltong beef on lower level; the one on the far left has temperature probe inserted.

Additional trials were performed to evaluate the contribution of each of the individual components of the marinade mixture on the various pathogens. In these trials, pathogen-inoculated beef was subjected to marination using no marinade components, salt alone, spice alone, vinegar alone, or the complete marinade. In the non-marinade control, salt-only, and spice-only marinades, water volume was added to the equivalent formulation volume of vinegar in the vinegar-alone or complete marinade versions to achieve the same level of liquid absorption on beef during vacuum tumbling prior to drying that may affect water activity during drying.

2.4. Drying of Marinaded Beef Pieces

Marinaded beef pieces were then hung in a humidity oven (Hotpack Model #435315, SP Industries, Warminster, PA, USA) set to 23.9 °C (75 °F) and 55% RH, using paper clips to hang beef pieces from bars set across the top and middle of the humidity chamber. Thermocouple temperature probes (×4) connected to a handheld temperature recorder were run into the chamber; 2 were used to record chamber temperature and 2 were inserted into 2 beef pieces to record internal beef temperature. An additional handheld humidity monitor was used to record both chamber humidity and temperature as well (Figure 2E,F).

2.5. Beef Sampling, Microbial Enumeration, Water Activity Testing, and Enterotoxin Detection

Each series of pathogen-inoculated trials was performed in duplicate trials with beef obtained from different animals. Within each trial, beef samples were tested in triplicate at all sampling points within each trial (initial inoculation, post-marinade, and after 2, 4, 6, 8, and 10 days of drying; $n = 6$). Each series of pathogen-inoculated trials included a set of uninoculated beef that was also subjected to biltong processing (marination, tumbling, and up to 10 days of drying). The uninoculated beef was used to insert temperature probes and for water activity (A_W) measurements throughout the drying process to ensure critical factors of temperature, humidity, and internal A_W were achieved and without fear of handling pathogen-inoculated pieces. Water activity was determined by slicing beef pieces in half and placing the innermost meat portion towards the humidity sensor in the sample chamber of the model HC2-AW-USB probe and analyzed using HW4-P-Quick software (Rotronic, Hauppauge, NY, USA). This provided Aw measurements for the inside of the thick beef pieces. The USDA-FSIS has indicated that vacuum-tumbled beef is considered 'non-intact beef' and internal A_W is a subject of concern for possible enterotoxin production by internalized *S. aureus* that might occur during vacuum-tumbling.

Acid-adapted cultures were used to desensitize the cultures to subsequent acidic treatment during processing and plated on TSA containing antibiotics instead of enumerating on selective/differential agars which are often known to be inhibitory to stressed cells [26]. Samples of beef retrieved from the humidity oven were placed in 6 × 9-inch Whirl-pak filter-stomacher bags (Nasco, Fort Atkinson, WI) to which 100 mL of neutralizing buffered peptone water (nBPW) was added. The stomacher bag was then stomached on a paddle masticator at high power for 2 min and subsequent dilutions were then made in 0.1% BPW. Select dilutions were then plated in duplicate on TSA + antibiotics for the respective pathogens described earlier and as previously used with Salmonella [10,26]. Plates were then incubated at 30 °C (*L. monocytogenes*) or 37 °C (*E. coli*, *S. aureus*) for 48 hrs before enumeration. At late stages of drying (i.e., ≥6 days), samples were often enumerated by plating 0.2 mL on each of 5 plates (i.e., 1 mL) to increase the sensitivity of detection at lower microbial levels.

Beef samples inoculated with a 4-strain mixture of *S. aureus* were sampled for enterotoxin detection. ELISA kits used for detecting SEA and SEB present in food, intestinal fluids, and liquid samples were obtained from Chondrex, Inc. (Woodinville, WA, USA). Test samples were analyzed according to the manufacture's protocol for *S aureus* strains producing SEA and SEB enterotoxin which included *S. aureus* ATCC 8095 (SEA), ATCC 13565 (SEA), ATCC 14458 (SEB), and ATCC 51740 (SEB). Testing included samples from each of 2 trials of biltong processing using the complete marinade: four samples tested for each duplicate trial replication at time points (a) after marination and (b) after drying for 10 days in the temperature-controlled humidity oven. Beef samples were extracted by addition of 100 mL of sterile water to filter-membrane bags and stomaching at high speed for 2 min in a paddle mixer as described earlier; recovered liquid samples were then centrifuged to remove debris and tested for enterotoxin levels. SEA and SEB enterotoxin standards (supplied with the kit) were prepared in a range of 0.16–10 ng/mL. The standards, samples, and detection antibody were diluted in sample/standard/detection antibody dilution buffer. The assay included addition of 50 µL of diluted standards, samples, and detection antibody to microplate wells precoated with primary antibody. The plates were then incubated for 1 hr on a plate shaker at room temperature (20 °C). After incubation, 100 µL of TMB colorimetric substrate (tetramethylbenzidine) was added to each well followed by an additional incubation for 25 min at room temperature on the plate shaker. After incubation the plates were washed 3 times using 1× wash buffer supplied with the kit. The reaction was then stopped by adding 50 µL of stop solution containing 2N sulfuric acid to each well. Plates were read at 450 nm using a GENios microplate reader (Tecan Inc, Morrisville, NC, USA) and analyzed with its associated Magellan software (ver. 7.1, Tecan).

2.6. Statistical Analysis

Each trial in this study was performed in duplicate replication with 3 samples tested per sampling period within each trial ($n = 6$) in accordance with validation testing criteria established by the NACMCF [28] and accepted by USDA-FSIS [29]. All replications were performed as autonomous and separate experiments using separately inoculated cultures and meat from different animals. Data are presented as the mean of multiple replications with standard deviation of the mean represented by error bars. Statistical analysis was performed using the statistics functions in Sigma-Plot ver. 13 (Systat Software, San Jose, CA, USA). Timed data series were statistically analyzed by repeated measures one-way analysis of variance (RM-ANOVA) using the Holm–Sidak test for pairwise multiple comparisons to determine significant differences ($p < 0.05$) between treatments. Data treatments with different letters are significantly different ($p < 0.05$); treatments with the same letter are not significantly different ($p > 0.05$).

3. Results and Discussion

3.1. Acid Adaptation of Bacterial Cultures

The bacterial cultures used in our study (*E. coli* O157:H7, *L. monocytogenes*, *S. aureus*) were cultured in TSB with varying concentrations of glucose (0%, 0.25%, 1.0%) to confirm their response to glucose concentration. As observed previously with *Salmonella* serovars [27], all three sets of strains produced correspondingly lower pH levels when incubated in TSB of increasing glucose concentration (Figure 3). For the purpose of ensuring that the biltong process is sufficiently robust to achieve high log reductions of even acid-adapted cultures, all strains used for biltong trials were grown in TSB containing 1% glucose.

Acid adaptation with *E. coli* O157:H7, *L. monocytogenes*, and *S. aureus* shows that they can lower the pH of the growth medium from approximately pH 7.0 (spent medium made without glucose) to as low as pH 4.4–4.8 in medium supplemented with 1% glucose (Figure 3). Acid adaptation was originally noted in *Salmonella* whereby growth in high glucose resulted in *Salmonella* being more tolerant of subsequent acidic conditions and has been shown to occur without genetic expression or protein synthesis whereby cells can modulate internal pH to a degree using membrane proton pumps. Other studies have noted a different process requiring genetic expression and protein synthesis, resulting in the Acid Tolerance Response (ATR) which is also 'adaptive' in the sense that genetic expression is often the result of responding to environmental stress conditions. These phenomena have been observed in *Salmonella*, shigatoxigenic *E. coli*, *L. monocytogenes*, and other pathogens [30,31]. Regardless of the exact mechanism, microbial acid adaption/tolerance/resistance is a concern for food processing that often rely on acidification via fermentation, food product formulation, or acidic antimicrobial interventions for food safety. The National Advisory Committee for the Microbiological Criteria for Foods (NACMCF) has recommended acid adaptation of cultures that are to be used in food challenge trials to ensure that food processes can inhibit even acid-tolerant organisms, and is preferred by US regulatory agencies (USDA-FSIS, FDA) for culture treatment when used in such studies. Bacterial cells accustomed to lower pH from their growth conditions should not be as sensitive to low pH conditions (acid treatment) and should therefore require a more robust process to result in significant microbial reductions. Investigators have shown that significant differences occur when comparisons are made between acid adapted and non-adapted cultures whereby non-adapted cultures decline faster/more when treated with an acidic treatment [30–32]. However, this issue is not clear since some investigators have demonstrated the opposite reaction (i.e., acid-adapted cultures showing greater reductions during acidic treatment). One explanation might be that these investigators have used selective/differential media (XLD/XLT4 for *Salmonella*, PALCAM/MOX for *L. monocytogenes*, SMAC for STEC *E. coli*) for enumeration of these pathogens; these media are harsh, and can inhibit stressed cells as would occur during acid-adaptation [33,34]. The enumeration of stressed cells showing significantly lower counts on selective/differential media has long been noted, and was recently confirmed in our comparison of enumeration of *Salmonella* serovars on TSA vs. HE vs. XLD after 4 different stresses [27].

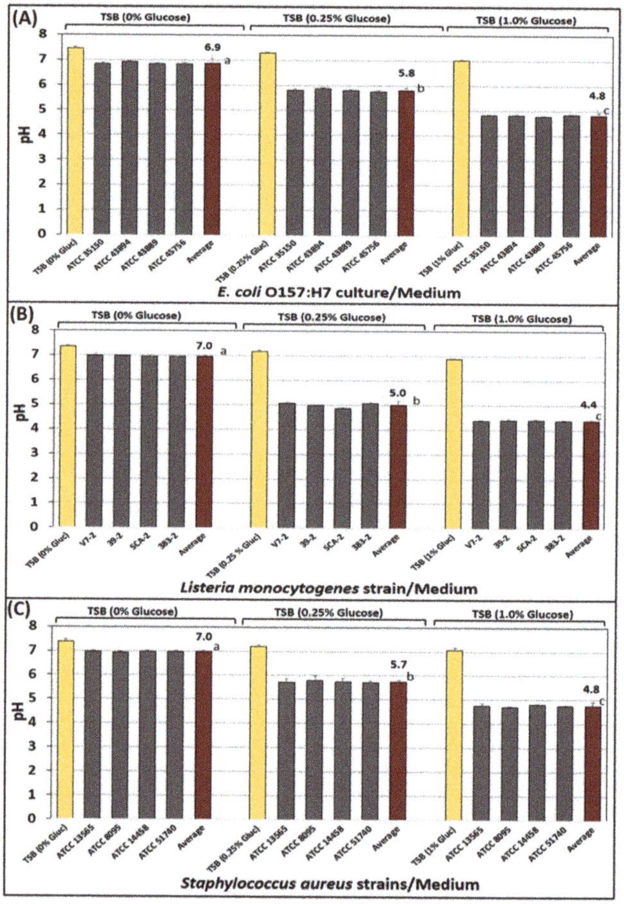

Figure 3. Response of various strains of *E. coli* O157:H7, *L. monocytogenes*, and *S. aureus* when grown in TSB medium at various glucose concentration (0%, 0.25%, 1.0%) at 37 °C (*S. aureus*, *E. coli* O157:H7) or 30 oC (*L. monocytogenes*): (**A**) *E. coli* O157:H7 (ATCC 35150, 41894, 43889, 45756); (**B**) *L. monocytogenes* (V7-2, 39-2, SCA-2, 383-2); (**C**) *S. aureus* (ATCC 8095, 13565, 14458, 51740). The values of the average pH for each set of 4 strains is listed above the average graph bar. Data are presented as the mean of triplicate replications and error bars represent the standard deviation from the mean. Means (for average pH) with different letters are significantly different as determined by one-way ANOVA using the Holm–Sidak test for pairwise multiple comparisons to determine significant differences ($p < 0.05$); means with different letters are significantly different ($p < 0.05$).

3.2. Biltong Process Reduction and Contribution of the Individual Components of the Marinade

Biltong inoculated with multi-strain acid-adapted cultures of *E. coli* O157:H7, *L. monocytogenes*, or *S. aureus* were subjected to a standardized biltong process described earlier (Figure 4). Data was obtained from 2 trials performed for each pathogen whereby triplicate samples were taken at each sampling point for each trial (post-inoculation; post-marinade, i.e., 0-day drying), and after 2, 4, 6, 8, and 10 days of drying at 23.9 °C (75 °F) and 55% RH. The data shows that a ≥5-log reduction was obtained for each pathogen, but not at the same time frame of processing. The 5-log reduction target was obtained for *E. coli* O157:H7 in 5 days (extrapolated), for *L. monocytogenes* in 6 days, and *S. aureus* in 7 days (extrapolated) (Figure 4).

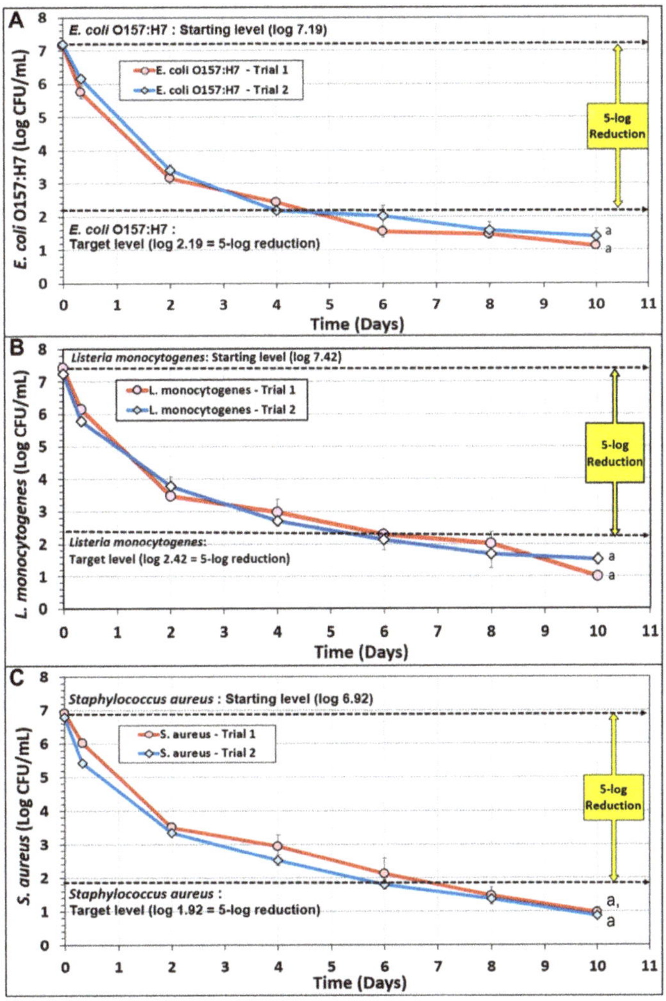

Figure 4. Biltong process reduction of beef inoculated with acid-adapted multi-strain cocktails of (**A**) *E. coli* O157:H7, (**B**) *L. monocytogenes*, and (**C**) *S. aureus*. The data represents 2 separate trials performed with triplicate sampling at each sampling point ($n = 6$). Duplicate trials with each pathogenic inoculum were analyzed by repeated measures one-way analysis of variance (RM-ANOVA) using the Holm–Sidak test for pairwise multiple comparisons to determine significant differences ($p < 0.05$) of treatment; treatments with the same letters are not significantly different ($p > 0.05$).

Another reason we have readily achieved the 5-log reduction target is that we inoculated samples with a fixed volume of inoculum giving a fixed population/sample (instead of dipping into inoculum) and we recover by stomaching samples in a fixed volume of diluent (100 mL) regardless of the dryness of the underlying beef. Other studies have enumerated on a cfu/gram basis whereby the remaining inoculum is concentrated as the underlying beef dries [20,35–37]. Our method eliminates this discrepancy since the underlying beef may lose up to 65% of its weight by 8–10 days of drying due to moisture loss [10].

The contribution of the individual components of biltong marinade mixtures to process lethality was examined during additional biltong trials. These additional trials included *E. coli* O157:H7-inoculated and *L. monocytogenes*-inoculated beef that were 'marinaded' with water-only, salt-only, and spice-only marinades (all in water), to equate the wetting of beef with vinegar during the complete or vinegar-only marination (Figure 5). The data shows that pathogen-inoculated beef (i.e., control/CTL) vacuum-tumbled with just added water, resulted in approximately 3.5–4.0 log reduction after 10 days of drying (Figures 5A and 6B). Pathogen-inoculated beef that was vacuum-tumbled with spices (in water) gave slightly higher levels of reduction, but not significantly different ($p > 0.05$) from controls (Figure 5). With salt (in water) and vinegar, some differences were observed depending on the pathogen. For instance, salt and vinegar showed similar processing effects during drying with *E. coli* O157:H7-inoculated beef, significantly different ($p < 0.05$) from both controls/water trials as well as the full complement of marinade components, but not significantly different ($p > 0.05$) from each other and both also demonstrated ≥5-log reduction of *E. coli* O157:H7 within 7 days (Figure 5A). Similar trials with salt-alone (in water) vs. vinegar-alone marination showed different results with *L. monocytogenes*. With *L. monocytogenes*-inoculated beef, salt (in water) marination gave a slightly greater reduction (but not significantly different) than for control and spice marination, yet vinegar marination gave even greater log reduction that was not significantly different from the full complement of marinade (Figure 5B). The control (CTL) data demonstrates that even without antimicrobial interventions that might be provided by marinade components, significant pathogen reduction (i.e., 3.5–4.0 log) still occurs during the drying process. This provides validation of >2-log reduction for those processes not demonstrating >5-log reduction of pathogen as one of the two alternative processes allowed for biltong processing by USDA-FSIS.

3.3. Biltong Process Temperature, Relative Humidity, and Water Activity (Aw) Measurements

Oven temperature measurements (2 probes) showed a range of approximately 23.9 ± 1.4 °C (75 ± 2.5 °F) for the 3 different pathogen trials and internal beef temperatures (2 probes) that gradually reached the temperature range of the oven (Figure 6). The RH probe was positioned near the midpoint of the oven near the far oven wall directly across from the wall that housed the oven fan (Figure 6). Based on the traces observed, slight tweaks were periodically made to the temperature and humidity controls during the drying period; however, these subtle tweaks could only be made after a period of observed traces since at any given time the temperature or humidity could be at a low or high point in the observed periodicity cycle.

Uninoculated beef was also processed during each of the pathogenic trials and used for A_W and temperature measurements so that handling of pathogen-inoculated beef would not be necessary for these measurements. Water activity measurements were closely aligned during all 3 pathogen trials whereby the treatments showed no significant differences by RM-ANOVA analyses (Figure 7).

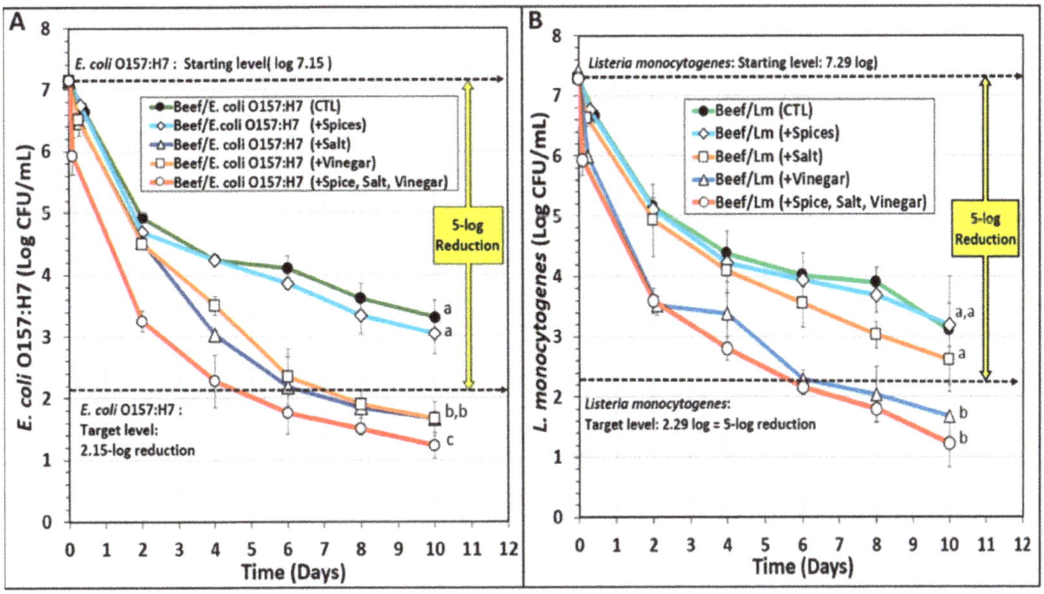

Figure 5. Comparison of individual marinade components and complete marinade on pathogen lethality during biltong processing of beef inoculated with (**A**) *E. coli* O157:H7 or (**B**) *L. monocytogenes*. The data represents an average of 2 separate trials performed with triplicate sampling at each sampling point; combined, $n = 6$ for each sampling period. Treatments with a particular pathogenic inoculum were analyzed by repeated measures one-way analysis of variance (RM-ANOVA) using the Holm–Sidak test for pairwise multiple comparisons to determine significant differences ($p < 0.05$) of treatment; treatments with the same letters are not significantly different ($p > 0.05$).

3.4. Analysis of Staphylococcal Enterotoxin A (SEA) and B (SEB) during Biltong Processing with S. aureus-Inoculated Beef

Strains of *S. aureus* used for beef inoculation included 2 strains that were known to be SEA-positive (ATCC 8095, ATCC 13565) and 2 known to be SEB-positive (ATCC 14458, ATCC 51740). Preliminary trials with the ELISA enterotoxin assay kit confirmed enterotoxin production by these strains (data not shown). Although our inoculum cells were centrifuged and resuspended, SEA and SEB were still detected from raw/inoculated post-marinaded beef indicative of residual carryover of SEA (0.91 ng/mL) and SEB (1.01 ng/mL) from the inoculum cell suspension. Perhaps additional washing of the cell pellets that were resuspended for use as biltong inoculum would have resulted in lower residual/initial levels on the inoculated beef. Subsequent samples obtained from biltong dried for 10 days demonstrated lower levels of SEA (0.43 ng/mL) and SEB (0.44 ng/mL) detected on beef after 10 days of drying (Figure 8). Both post-marinade and 10-day dried beef were subject to sample recovery in the same volume of extraction diluent and not on a per gram basis that would otherwise cause quantification analyses to be affected by degree of moisture loss (similar to how our microbial enumerations were performed). The data demonstrates that SEA and SEB detected after 10 days of biltong process drying were 53% and 57% lower than the levels detected at Day 0 (post-inoculation, rinsing, and marination) suggesting that enterotoxin production did not occur during our biltong process.

Because of the thickness of biltong beef, vacuum tumbling raises a concern as 'non-intact beef' (a safety concern for USDA-FSIS) that could have issues related to internalization of *S. aureus* that might result in production of staphylococcal enterotoxin if growth and conditions allowed. Our prior work showed that biltong made with 2.2% NaCl in the total biltong marinade formulation (including the weight of beef) approaches a moisture loss of ~60%, an internal A_W of 0.82 (A_W of 0.69 for ground biltong), a biltong pH of ~5.26 and 3–4% NaCl in the resulting biltong after 8 days of drying [9,10]. These conditions individually are not conducive to staphylococcal enterotoxin production, and less so when they occur simultaneously [29–32]. This level of water activity, though safe for enterotoxin production, may still allow mold growth which is aesthetically unappealing and reflects negatively on the product and company. Processors can do various things to mitigate mold growth such as the use of 'single pack' servings that are expected to be consumed once opened, the addition of moisture/oxygen absorbing packets, and/or the use of mold inhibitors in the marinade formulation.

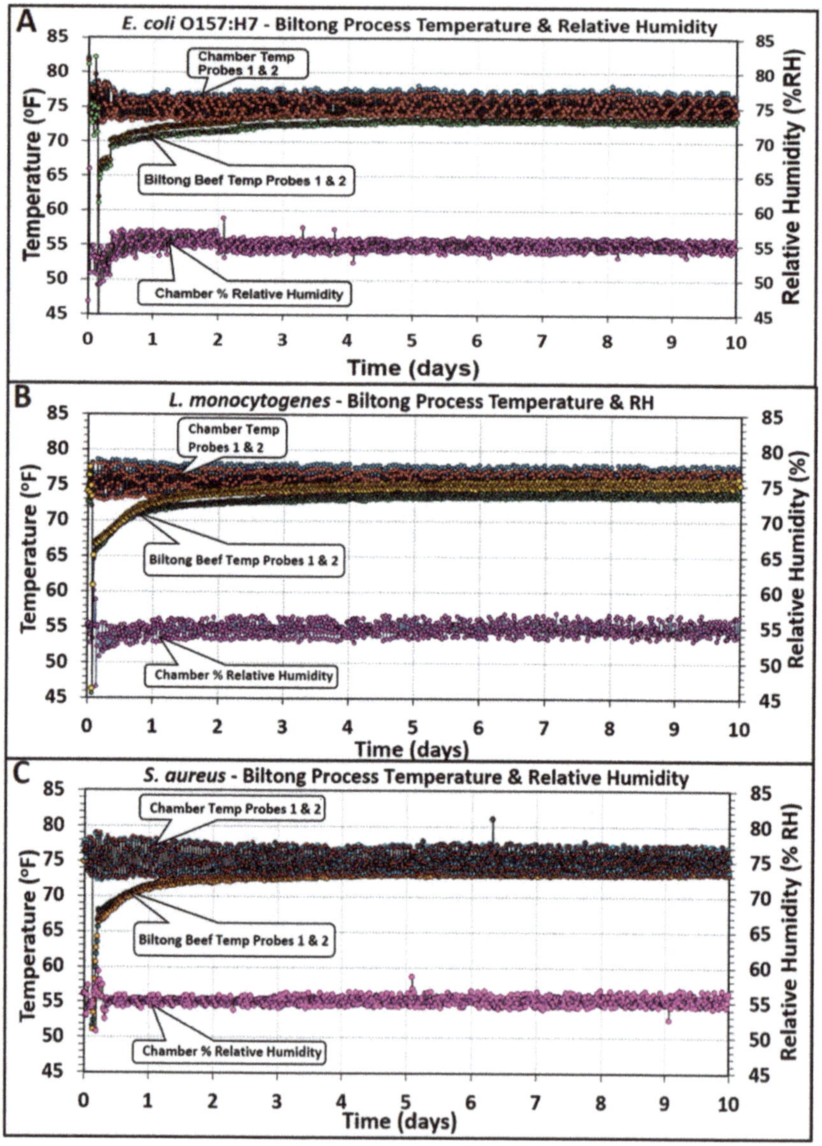

Figure 6. Temperature (°F) and relative humidity (% RH) oven conditions with (**A**) *E. coli* O157:H7, (**B**) *L. monocytogenes*, and (**C**) *S. aureus* inoculated beef during biltong processing. The temperature and humidity range observed were 75 ± 2.5 °F (23.9 ± 1.4 °C) and 55% ± 1.5% RH, respectively.

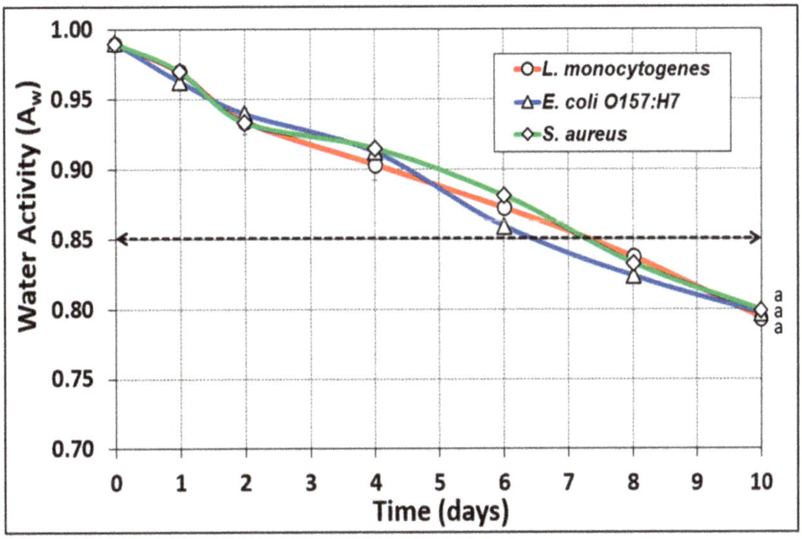

Figure 7. Water activity (A_W) from interior portions of biltong beef during trials with *E. coli* O157:H7, *L. monocytogenes*, or *S. aureus*, held at 23.9 °C (75 °F) and 55% RH for up to 10 days. Repeated measures one-way analysis of variance (RM-ANOVA) using the Holm—-Sidak test for pairwise multiple comparisons to determine significant differences ($p < 0.05$); treatments with the same letter are not significantly different ($p > 0.05$).

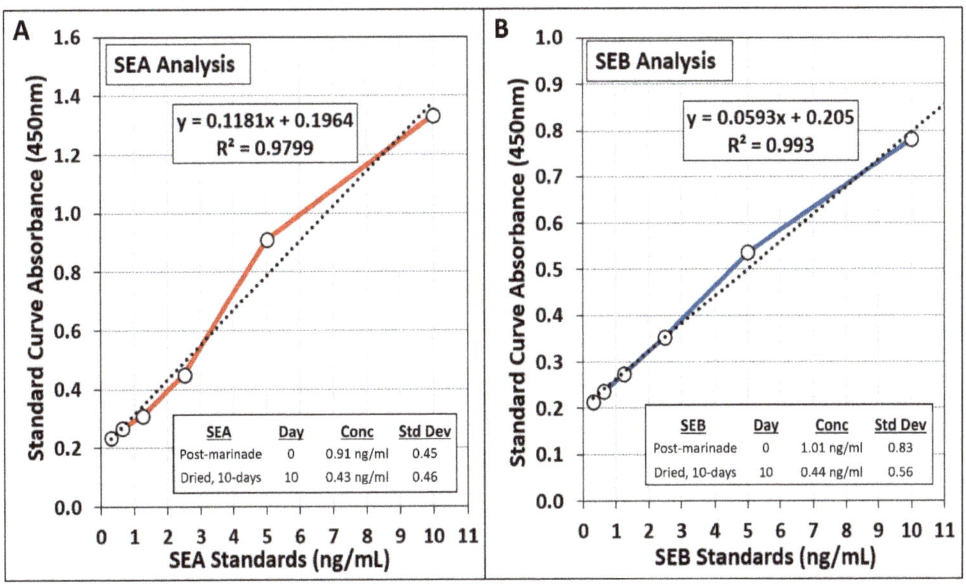

Figure 8. Analysis of (**A**) SEA and (**B**) SEB from *S. aureus*-inoculated biltong beef after rinse/marinade treatment (Day 0, initial measurement) and after 10 days of drying (latter measurement) using staphylococcal enterotoxin ELISA quantification kit. Graphs show standard curve plot and SEA and SEB analyses of inoculated samples post-marinade and post-dried (10 day) samples. Beef sample enterotoxin data levels were derived from assay absorbance and quantified from the equation obtained for the standard curve trend line ($n = 4$).

3.5. Comparison of Microbial Reduction of E. coli O157:H7, L. monocytogenes, S. aureus, and Salmonella serovars during Biltong Processing

The data presented herein demonstrate that a basic biltong process and marinade composed of spices (pepper, coriander), salt, and vinegar, applied during vacuum tumbling for 30 min and followed by drying at 23.9 °C (75 °F) and 55% RH can result in ≥5-log reduction (in 5–7 days) with *Salmonella* serovars [9,10], as well as *E. coli* O157:H7, *L. monocytogenes*, and *S. aureus* (this study) (Figure 9). The 4 major groups of foodborne pathogens that could be associated with raw beef and potentially a concern for survival during the biltong process react similarly when subjected to the basic biltong process used in this study (Figure 9).

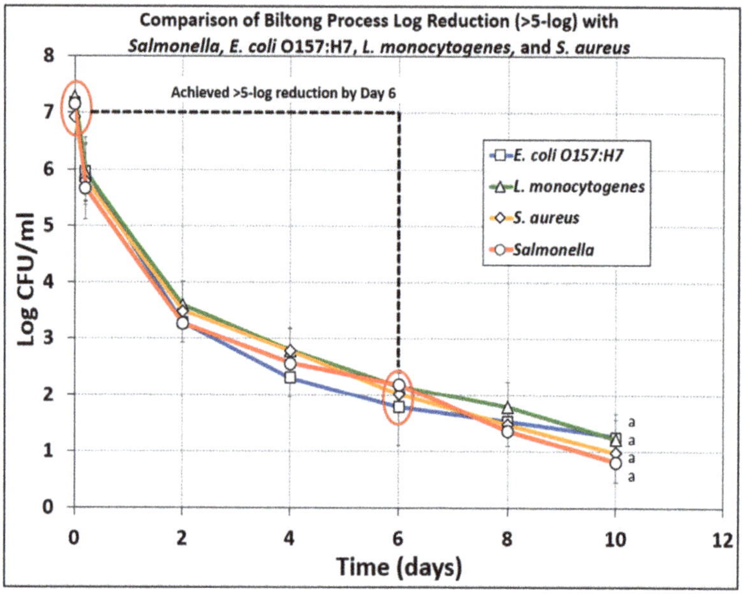

Figure 9. Comparison microbial lethality of a basic biltong process for >5-log reduction of *Salmonella* (adapted from [10]), *E. coli* O157:H7, *L. monocytogenes*, and *S. aureus* (this study). The data represent the average of 2 separate trials, each performed with triplicate sampling at each sampling point (combined, $n = 6$ for each sampling period). Treatments were analyzed by repeated measures one-way analysis of variance (RM-ANOVA) using the Holm–Sidak test for pairwise multiple comparisons to determine significant differences ($p < 0.05$) of treatment; treatments with the same letters are not significantly different ($p > 0.05$).

4. Conclusions

The data provided in our study show that a basic biltong process can achieve ≥5-log reduction and equivalent lethality among four major groups of foodborne pathogens associated with raw beef to provide a safe dried beef protein snack food. We feel the data presented herein should provide sufficient validation for USDA-FSIS process approval and product acceptance by food safety managers of retail supermarket chains, given their inter in ensuring safe foods for consumers.

Author Contributions: Conceptualization, P.M.M.; methodology, K.G. and C.K.; software, P.M.M.; validation, K.G., C.K. and P.M.M.; formal analysis, K.G. and C.K.; investigation, K.G. and C.K.; resources, P.M.M.; data curation, P.M.M.; writing—original draft preparation, K.G.; writing—review and editing, P.M.M.; visualization, P.M.M.; supervision, P.M.M.; project administration, P.M.M.; funding acquisition, P.M.M. All authors have read and agreed to the published version of the manuscript.

Funding: Funding for the study was contributed in part by the Robert M. Kerr Food & Ag Products Center, the Advance Foods—SE Gilliland Professorship in Microbial Food Safety (#21-57200), the USDA National Institute of Food and Agriculture (Hatch Project #OKL03090), the Oklahoma Agricultural Experiment Station, the Division of Agricultural Sciences and Natural Resources at Oklahoma State University, and various industry sources.

Institutional Review Board Statement: The work performed in this study was performed under Protocol #17-8 approved by the OSU Institutional Biosafety Committee within the Office of University Research Compliance.

Informed Consent Statement: Not applicable.

Data Availability Statement: Not applicable.

Acknowledgments: The authors would like to acknowledge Jake Nelson, M.S. Meat Science/FAPC, and Meryl Silverman (Senior Food Technologist, Office of Policy and Program Development, Food Safety and Inspection Service, USDA, Washington, DC, USA) for their technical assistance in meat processing and on USDA-FSIS regulatory policy with respect to dried beef, respectively.

Conflicts of Interest: The authors declare no conflict of interest. The funders had no role in the design of the study; in the collection, analyses, or interpretation of data; in the writing of the manuscript, or in the decision to publish the results.

References

1. Wentworth, E.N. Dried meat: Early man's travel ration. *Agric. Hist.* **1956**, *30*, 2–10.
2. Bermúdez-Aguirre, D.; Tapia, M.S.; Welti-Chanes, J. *Specialty Foods*; DEStech Publications, Inc.: Lancaster, PA, USA, 2008; pp. 447–461.
3. Challa, H.J.; Bandlamudi, M.; Uppaluri, K.R. Paleolithic Diet. In *StatPearls*; StatPearls Publishing, LLC.: Treasure Island, FL, USA, 2021.
4. Kussaga, J.B.; Jacxsens, L.; Tiisekwa, B.P.; Luning, P.A. Food safety management systems performance in African food processing companies: A review of deficiencies and possible improvement strategies. *J. Sci. Food Agric.* **2014**, *94*, 2154–2169. [CrossRef]
5. Burfoot, D.; Everis, L.; Mulvey, L.; Wood, A.; Betts, R. Literature Review on Microbiological Hazards Associated with Biltong and Similar Dried Meats. Available online: https://www.food.gov.uk/sites/default/files/media/document/574-1-1007_B13015_Final_Report.pdf (accessed on 13 March 2022).
6. BC-CDC. South African RTE dehydrated meats (biltong, chili bites, droëwors). In *Food Issue, Notes from the Field*; British Columbia Centre for Disease Control: Vancouver, BC, Canada, 2020; pp. 1–7.
7. USDA-FSIS. *FSIS Compliance Guideline for Meat and Poultry Jerky Produced by Small and very Small Establishments*; U.S. Food Safety and Inspection Service: Washington, DC, USA, 2014; pp. 1–54.
8. USDA-FSIS. *Salmonella Compliance Guidelines for Small and very Small Meat and Poultry Establishments That Produce Ready-To-Eat (RTE) Products and Revised Appendix A, June 2017*; U.S. Food Safety and Inspection Service: Washington, DC, USA, 2017; pp. 1–37.
9. Karolenko, C.; Muriana, P. Quantification of process lethality (5-log reduction) of *Salmonella* and salt concentration during sodium replacement in biltong marinade. *Foods* **2020**, *9*, 1570. [CrossRef] [PubMed]
10. Karolenko, C.E.; Bhusal, A.; Nelson, J.L.; Muriana, P.M. Processing of biltong (dried beef) to achieve USDA-FSIS 5-log reduction of *Salmonella* without a heat lethality step. *Microorganisms* **2020**, *8*, 791. [CrossRef] [PubMed]
11. Nickelson, R.; Luchansky, J.B.; Kaspar, C.; Johnson, E. *Dry Fermented Sausage and Escherichia coli O157:H7 Validation Research. An Executive Summary Prepared by The Blue Ribbon Task Force of the National Cattlemen's Beef Association*; Research Report No. 11-316.; NCBA: Chicago, IL, USA, 1996.
12. Matsheka, M.I.; Mpuchane, S.; Gashe, B.A.; Allotey, J.; Khonga, E.B.; Coetzee, S.H.; Murindamombe, G. Microbial quality assessment and predominant microorganism of biltong produced in butcheries in Gaborone, Botswana. *Food Nutr. Sci.* **2014**, *5*, 11. [CrossRef]
13. Naidoo, K.; Lindsay, D. Pathogens associated with biltong product and their in vitro survival of hurdles used during production. *Food Prot. Trends* **2010**, *30*, 532–538.
14. USDA-FSIS. Food Science Research Priorities; Data Gaps. Available online: https://www.fsis.usda.gov/science-data/research-priorities (accessed on 5 March 2022).
15. Laufer, A.S.; Grass, J.; Holt, K.; Whichard, J.M.; Griffin, P.M.; Gould, L.H. Outbreaks of *Salmonella* infections attributed to beef –United States, 1973-2011. *Epidemiol. Infect.* **2015**, *143*, 2003–2013. [CrossRef]
16. CDC. Outbreak of Salmonellosis associated with beef jerky–New Mexico, 1995. *MMWR Morb. Mortal. Wkly. Rep.* **1995**, *44*, 785–788.
17. Eidson, M.; Swell, C.M.; Graves, G.; Olson, R. Beef jerky gastroenteritis outbreaks. *J. Environ. Health* **2000**, *62*, 9.
18. Omer, M.K.; Álvarez-Ordoñez, A.; Prieto, M.; Skjerve, E.; Asehun, T.; Alvseike, O.A. A systematic review of bacterial foodborne outbreaks related to red meat and meat products. *Foodborne Pathog. Dis.* **2018**, *15*, 598–611. [CrossRef]

19. Mhlambi, S.G.; Naidoo, K.; Lindsay, D. Enterotoxin-producing *Staphylococcus* strains associated with South African biltong at point of sale. *J. Food Safety* **2010**, *30*, 307–317. [CrossRef]
20. Naidoo, K.; Lindsay, D. Survival of *Listeria monocytogenes*, and enterotoxin-producing *Staphylococcus aureus* and *Staphlyococcus pasteruri*, during two types of biltong-manufacturing process. *Food Cont.* **2010**, *21*, 1042–1050. [CrossRef]
21. Muriana, P.M.; Quimby, W.; Davidson, C.A.; Grooms, J. Postpackage pasteurization of ready-to-eat deli meats by submersion heating for reduction of *Listeria monocytogenes*. *J. Food Prot.* **2002**, *65*, 963–969. [CrossRef] [PubMed]
22. Muriana, P.M.; Eager, J.; Wellings, B.; Morgan, B.; Nelson, J.; Kushwaha, K. Evaluation of antimicrobial interventions against *E coli* O157:H7 on the surface of raw beef to reduce bacterial translocation during blade tenderization. *Foods* **2019**, *8*, 80. [CrossRef] [PubMed]
23. Berry, E.D.; Cutter, C.N. Effects of acid adaptation of *Escherichia coli* O157:H7 on efficacy of acetic acid spray washes to decontaminate beef carcass tissue. *Appl. Environ. Microbiol.* **2000**, *66*, 1493–1498. [CrossRef]
24. Panneerseelan, L.; Muriana, P.M. An immunomagnetic PCR signal amplification assay for sensitive detection of *Staphylococcus aureus* enterotoxins in foods. *J. Food Prot.* **2009**, *72*, 2538–2546. [CrossRef]
25. Mehrotra, M.; Wang, G.; Johnson, W.M. Multiplex PCR for detection of genes for *Staphylococcus aureus* enterotoxins, exfoliative toxins, toxic shock syndrome toxin 1, and methicillin resistance. *J. Clin. Microbiol.* **2000**, *38*, 1032–1035. [CrossRef]
26. Wilde, S.; Jørgensen, F.; Campbell, A.; Rowbury, R.; Humphrey, T. Growth of *Salmonella enterica* Serovar Enteritidis PT4 in media containing glucose results in enhanced RpoS-independent heat and acid tolerance but does not affect the ability to survive air-drying on surfaces. *Food Microbiol* **2000**, *17*, 679–686. [CrossRef]
27. Karolenko, C.E.; Bhusal, A.; Gautam, D.; Muriana, P.M. Selenite cystine agar for enumeration of inoculated *Salmonella* serovars recovered from stressful conditions during antimicrobial validation studies. *Microorganisms* **2020**, *8*, 338. [CrossRef]
28. NACMCF. Parameters for determining inoculated pack/challenge study protocols (National Advisory Committee on the Microbiological Criteria for Foods). *J. Food Prot.* **2010**, *73*, 140–202. [CrossRef]
29. USDA-FSIS. *FSIS Compliance Guideline HACCP Systems Validation (April 2015)*; USDA-FSIS, Ed.; USDA: Washington, DC, USA, 2015; pp. 1–68.
30. Álvarez-Ordóñez, A.; Prieto, M.; Bernardo, A.; Hill, C.; López, M. The acid tolerance response of *Salmonella* spp.: An adaptive strategy to survive in stressful environments prevailing in foods and the host. *Food Res. Intl.* **2012**, *45*, 482–492. [CrossRef]
31. Alvarez-Ordóñez, A.; Fernández, A.; Bernardo, A.; López, M. A comparative study of thermal and acid inactivation kinetics in fruit juices of *Salmonella enterica* serovar Typhimurium and *Salmonella enterica* serovar Senftenberg grown at acidic conditions *Foodborne Pathog. Dis.* **2009**, *6*, 1147–1155. [CrossRef] [PubMed]
32. Samelis, J.; Sofos, J.N.; Kendall, P.A.; Smith, G.C. Effect of acid adaptation on survival of *Escherichia coli* O157:H7 in meat decontamination washing fluids and potential effects of organic acid interventions on the microbial ecology of the meat plant environment. *J. Food Prot.* **2002**, *65*, 33–40. [CrossRef]
33. Calicioglu, M.; Sofos, J.N.; Samelis, J.; Kendall, P.A.; Smith, G.C. Inactivation of acid-adapted and nonadapted *Escherichia coli* O157:H7 during drying and storage of beef jerky treated with different marinades. *J. Food Prot.* **2002**, *65*, 1394–1405. [CrossRef]
34. Calicioglu, M.; Sofos, J.N.; Samelis, J.; Kendall, P.A.; Smith, G.C. Effect of acid adaptation on inactivation of *Salmonella* during drying and storage of beef jerky treated with marinades. *Int. J. Food Microbiol.* **2003**, *89*, 51–65. [CrossRef]
35. Jones, M.; Arnaud, E.; Gouws, P.; Hoffman, L.C. Effects of the addition of vinegar, weight loss and packaging method on the physicochemical properties and microbiological profile of biltong. *Meat Sci.* **2019**, *156*, 214–221. [CrossRef]
36. Burnham, G.M.; Hanson, D.J.; Koshick, C.M.; Ingham, S.C. Death of *Salmonella* serovars, *Escherichia coli* O157: H7, *Staphylococcus aureus* and *Listeria monocytogenes* during the drying of meat: A case study using Biltong and Droewors. *J. Food Safety* **2008**, *28*, 198–209. [CrossRef]
37. Petit, T.; Caro, Y.; Petit, A.S.; Santchurn, S.J.; Collignan, A. Physicochemical and microbiological characteristics of biltong, a traditional salted dried meat of South Africa. *Meat Sci.* **2014**, *96*, 1313–1317. [CrossRef]

microorganisms

Article

Genomic Characterization of *Escherichia coli* O8 Strains Producing Shiga Toxin 2l Subtype

Xi Yang [1], Qian Liu [1], Hui Sun [1], Yanwen Xiong [1], Andreas Matussek [2,3,4] and Xiangning Bai [1,2,3,*]

1 State Key Laboratory of Infectious Disease Prevention and Control, National Institute for Communicable Disease Control and Prevention, Chinese Center for Disease Control and Prevention, Beijing 102206, China; yangxicdc163@163.com (X.Y.); qianliu9608@gmail.com (Q.L.); sunhui@icdc.cn (H.S.); xiongyanwen@icdc.cn (Y.X.)
2 Division of Clinical Microbiology, Department of Laboratory Medicine, Karolinska Institute, 141 52 Stockholm, Sweden; anmatu@ous-hf.no
3 Division of Laboratory Medicine, Oslo University Hospital, 0372 Oslo, Norway
4 Division of Laboratory, Medicine Institute of Clinical Medicine, University of Oslo, 0372 Oslo, Norway
* Correspondence: xiangning.bai@ki.se

Abstract: Shiga toxin-producing *Escherichia coli* (STEC) can cause diseases ranging from mild diarrhea to fatal extra-intestinal hemolytic uremic syndrome (HUS). Shiga toxin (Stx) is the key virulence factor in STEC, two Stx types (Stx1 and Stx2) and several subtypes varying in sequences, toxicity, and host specificity have been identified. Stx2l is a newly-designated subtype related to human disease but lacks thorough characterization. Here, we identified Stx2l from five STEC strains (Stx2l-STECs) recovered from raw mutton and beef in China. Whole-genome sequencing (WGS) was used to characterize the Stx2l-STECs in this study together with Stx2l-STECs retrieved from public databases. Our study revealed that all the analyzed Stx2l-STEC strains belonged to the same serogroup O8. Multilocus sequencing typing (MLST) showed two sequence types (ST88 and ST23) among these strains. Stx2l-converting prophages from different sources shared a highly similar structure and sequence. Single-nucleotide polymorphism (SNP)-based analysis revealed genetic relatedness between the human-derived and food-derived strains belonging to ST23. To conclude, our study supported the designation of Stx2l and demonstrated diverse host range and geographical distribution of Stx2l-STECs.Stx2l-STEC strains from different sources showed a high genetic similarity with an identical O8 serogroup. Further studies are needed to investigate the epidemiological trait and pathogenic potential of Stx2l-STEC strains.

Keywords: Shiga toxin; *Escherichia coli*; Stx2l; whole genome sequencing

Citation: Yang, X.; Liu, Q.; Sun, H.; Xiong, Y.; Matussek, A.; Bai, X. Genomic Characterization of *Escherichia coli* O8 Strains Producing Shiga Toxin 2l Subtype. *Microorganisms* **2022**, *10*, 1245. https://doi.org/10.3390/microorganisms10061245

Academic Editor: Rodney A. Moxley

Received: 28 May 2022
Accepted: 17 June 2022
Published: 17 June 2022

1. Introduction

Shiga toxin-producing *Escherichia coli* (STEC) is a significant foodborne pathogen that can cause human gastrointestinal diseases ranging from mild diarrhea to life-threatening hemolytic uremic syndrome (HUS) [1,2]. The key virulence factor of STEC is Shiga toxin (Stx), which is encoded by *stx* located downstream of lysogenized lambdoid prophages [3,4]. Stx-converting prophages are highly mobile genetic elements that play an important role in horizontal gene transfer and STEC pathogenesis [5]. Stx comprises two immunologically distinct types (Stx1 and Stx2) [6]. Based on variations in amino acid sequences, a standardized taxonomic nomenclature proposed by Scheutz et al. distinguished Stx1 and Stx2 into various subtypes, i.e., three Stx1 (Stx1a, Stx1c, and Stx1d) and seven Stx2 (Stx2a to Stx2g) subtypes [7]. Different Stx subtypes vary in biologic activity leading to a difference in epidemiological association with patient outcomes. Stx2-producing strains are more often associated with severe disease than Stx1-producing strains [8,9]. Since the establishment of the standardized Stx subtyping approach, several novel Stx2 subtypes have been reported, including Stx2h to Stx2m and Stx2o [10–14]. The newly proposed Stx2l subtype, initially designed as Stx2e, has been identified in a few clinical and sheep isolates [9,15]. However, the characteristics of the Stx2l-STEC strains have been poorly elucidated.

In the present study, we identified the Stx2l subtype from raw meats-derived STEC strains in China and strains of diverse origins from other countries. Whole-genome sequencing (WGS) was performed to characterize the genomic features of Stx2l-STEC strains and Stx2l-converting prophages. The phylogenetic relatedness of the Stx2l-STEC strains in this study and those reported from other sources was assessed.

2. Materials and Methods

2.1. Ethics Statement

The current study was reviewed and approved by the ethics committee of the National Institute for Communicable Diseases Control and Prevention, China CDC, with the number ICDC-2017003.

2.2. Identification of Stx2l Subtype from STEC Strains

Identification of STEC strains carrying *stx2l* subtype (Stx2l-STEC) from our STEC collection was performed by an in-house *stx*_subtyping approach based on WGS data as previously described [16,17]. Briefly, an in-house *stx*_subtyping database was created including representative nucleotide sequences of all identified Stx1 and Stx2 subtypes. Notably, the originally-designated Stx2e variant Stx2e-O8-FHI-1106-1092 (AM904726.1) was redesignated as Stx2l [9]. Stx2l-STEC were identified by screening genome assemblies against the updated *stx*_subtyping database using ABRicate version 0.8.10 (https://github.com/tseemann/abricate) (accessed on 1 April 2022). Other Stx2l-STEC genomes were retrieved from the National Center for Biotechnology Information (NCBI, https://www.ncbi.nlm.nih.gov/) and EnteroBase databases (https://enterobase.warwick.ac.uk/species/ecoli) (all accessed on 1 April 2022).

To validate the taxonomic position of Stx2l, the full nucleotide sequences of Stx2l in this study were extracted from the genome assemblies; several representative nucleotide sequences of all the Stx2 subtypes (Stx2a to Stx2m and Stx2o) and variants were selected as references and downloaded from the GenBank. The amino acid sequences for the combined A and B subunits of Stx2 holotoxin were translated from the open reading frames. Phylogenetic trees based on the holotoxin amino acid sequences were reconstructed with three algorithms, Neighbor-Joining, Maximum Likelihood, and Maximum Parsimony, using MEGA 11 software (www.megasoftware.net) (accessed on 5 April 2022); the stability of the groupings was estimated by bootstrap analysis (1000 replications).

2.3. WGS-Based Molecular Characterization of Stx2l-STEC Strains

In silico serotyping was conducted by comparing genome assemblies against the SerotypeFinder database (https://cge.food.dtu.dk/services/SerotypeFinder/) (accessed on 10 April 2022) using ABRicate version 0.8.10 with default parameters. Multilocus sequence typing (MLST) was performed in silico through the online tool provided by the Warwick *E. coli* MLST scheme website (https://enterobase.warwick.ac.uk/species/ecoli/allele_st_search) (accessed on 10 April 2022).

2.4. Genomic Characterization of Stx2l-Converting Prophages

Complete genomes of the Stx2l-STEC strains were uploaded to the PHAge Search Tool Enhanced Release (PHASTER, http://phaster.ca/) (accessed on 12 April 2022) to identify Stx-converting phages. The intact Stx2l prophage sequences were extracted from the complete STEC genomes. To obtain the Stx2l prophages from the draft genomes, the RAST server [18] was used to annotate the draft Stx2l-STEC genomes; the sequences of the Stx2l-converting prophages from those draft genomes were reconstructed from multiple contigs based on BLASTn searching, RAST annotation, and progressive Mauve alignment [19]. Intact Stx2l prophages were defined when complete prophage structures were identified.

The gene adjacent to the integrase gene was designated as the phage insertion site [4]. Stx2l-converting prophages were compared and visualized using Easyfig [20].

2.5. Single Nucleotide Polymorphism (SNP)-Based Phylogeny of Stx2l-STECs

A whole-genome SNP phylogeny was used to assess the genomic relationship of the Stx2l-STEC strains reported so far. The core alignment of the SNPs was obtained by using snippy-multi in Snippy version 4.3.6 (https://github.com/tseemann/snippy) (accessed on 15 April 2022) with the default parameters; the Stx2l-STEC strain STEC306 in this study (SAMN21841557) was used as a reference. Gubbins version 2.3.4 [21] was then used to remove recombination from core SNP alignments and construct a maximum likelihood tree based on the filtered SNP alignments. The SNP distances were executed using snp-dists v0.7.0. (https://github.com/tseemann/snp-dists) (accessed on 15 April 2022). The visualization and annotation of the phylogenetic tree were produced by the web-tool ChiPlot (https://www.chiplot.online/#Phylogenetic-Tree (accessed on 18 April 2022).

2.6. Data Availability

The genome data presented in this study are publicly available in NCBI with the accession numbers SAMN20824184, SAMN21841566, SAMN21841557, SAMN21841558, SAMN21841559, SAMN16454166, SAMN12290412, SAMEA2593965 and SAMN05171628.

3. Results

3.1. Identification of Stx2l Subtype in Food-Derived STEC Strains

The in-house *stx* subtyping approach showed that the *stx2* sequences of five isolates in our STEC strains collection (n = 882) shared 99.68% nucleic acid sequence identity with the Stx2l subtype. Three strains were isolated from raw mutton samples collected in a market in Beijing, China, in December 2013. Two strains were isolated from one raw beef and one raw mutton sample collected in the same market in March 2014. These isolates were previously identified as Stx2e-STEC strains based on the nomenclature proposed in 2012 [7]. We performed BLASTn searching using the *stx2* reference sequence (AM904726.1) against the Refseq Genome Database (taxid:562); *stx2* sequences from six STEC genomes yielded a nucleic acid sequence identity above 99% with the reference *stx2l* sequence. In addition to the five raw meat-derived strains in this study, two food-derived STEC strains from the USA, and one gastroenteritis patient-derived STEC strain from Norway were found to carry *stx2l*. One human-derived Stx2l-STEC strain from the UK was found in the EnteroBase database. Thus, nine Stx2l-STEC genomes were included in the subsequent analysis.

Phylogenetic trees based on the holotoxin amino acid sequences of all Stx2 subtypes using the Neighbor-Joining (Figure 1), Maximum-Likelihood, and Maximum Parsimony algorithms shared the same topology (data not shown); the Stx2 of the five strains in this study and four strains from other sources clustered with Stx2l type sequence Stx2l-O8-FHI-1106-1092, while they formed a distinct lineage from other Stx2 subtypes (Figure 1). These data supported that these STEC strains harbor the *stx2l* subtype. Seven out of the nine strains shared identical Stx2l amino acid sequence, while one and two amino acid differences were found in the Stx2l representative strain FHI-1106-1092 and strain 879916 from food, respectively, when compared with the others. Identical IS2 family transposase was inserted into the intergenic region between the A and B subunits of Stx2 from three Stx2l-STEC isolates (STEC306, STEC307, and STEC308); the intergenic regions among other six *stx2l* sequences were identical, which contained 11 nucleotides (aggagttaagt).

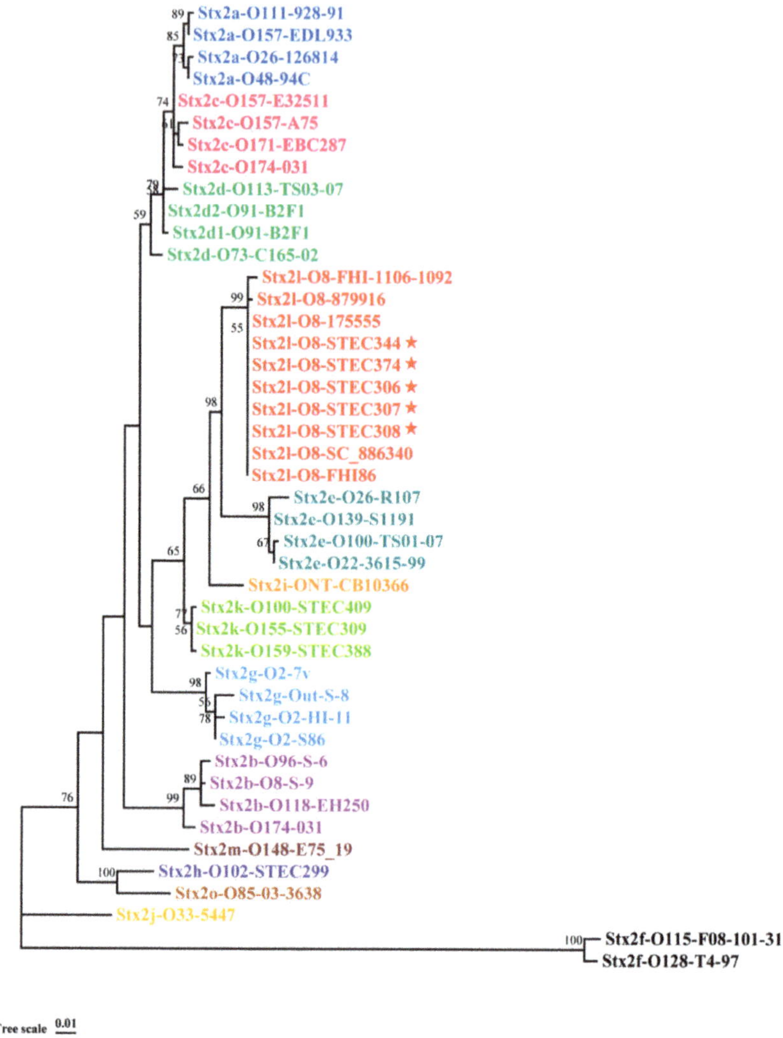

Figure 1. Phylogenetic tree of the Stx2 subtypes by the Neighbor-Joining method. The Neighbor-Joining tree was inferred from comparison of the combined A and B holotoxin amino acid sequences of all Stx2 subtypes. The numbers on the tree indicate the bootstrap values calculated for 1000 subsets for branch points > 50%. Tree scale, 0.01 substitutions per site. Stx2 subtypes are indicated by different colors. The red stars indicate strains in this study.

3.2. Stx2l-STEC Strains Belonged to Serogroup O8

The WGS data analysis revealed an identical O8 serogroup of the nine Stx2l-STEC strains. Two meat-derived Stx2l-STEC strains in this study, two food-derived strains from the USA, and one gastroenteritis patient strain from Norway were assigned to the same serotype O8:H30, while three meat-derived Stx2l-STEC strains in this study and one human-derived strain from the UK were assigned as O8:H9. Of note, all food-derived strains belonged to the same MLST type ST88, while two human strains belonged to the MLST type ST23.

3.3. Genetic Feature of Stx2l-Converting Prophages

The BLASTn search showed that the Stx2l-STEC strain STEC306 (SAMN21841557) harbored the most similar phage sequence with the other Stx2l prophages; thus, this complete Stx2l-converting prophage was selected as a reference to identify prophages from the draft genomes. Seven complete and two incomplete Stx2l prophages were obtained. The comparison of the Stx2l prophages demonstrated the genetic similarity among these prophages from different sources (Figure 2). Similar to other Stx prophages [22], three major modules, i.e., the integration cassette, lysis cassette, and morphogenetic related genes, were found in the Stx2l-converting prophages; the sequence and structure of the three modules each were nearly identical among the different Stx2l-converting prophages. Variability was observed at their insertion sites and phage integrase; Stx2l prophages of the four food-derived strains were inserted at *ypjA* (adhesin-like autotransporter YpjA/EhaD), while the human-derived Stx2l prophages and three food-derived Stx2l prophages were inserted at *parB* (ParB/RepB/Spo0J family partition protein). In addition to the identical integrase (WP_033813161.1) possessed by all Stx2l prophages, the recombinase family protein (WP_000135615.1) and integrase domain-containing protein (WP_248232026.1) were found in the human-derived Stx2l prophages and other three food-derived Stx2l prophages.

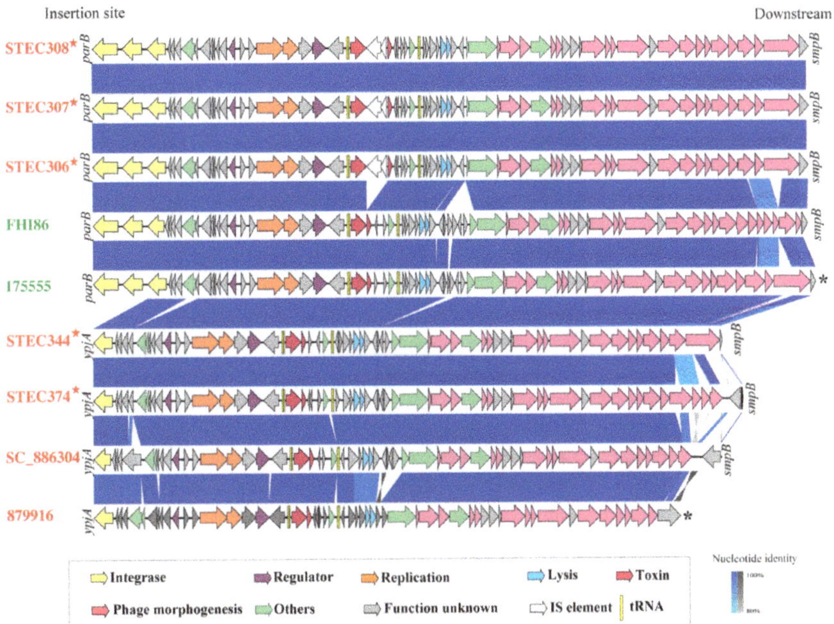

Figure 2. Comparison of the nine Stx2l-converting prophages. The arrows indicate gene directions. The coding sequences are represented by arrows and linked by blue bars shaded to represent the nucleotide identity and gray bars shaded to represent the nucleotide identity of reverse complemented homologous sequences, as indicated in the legend. The color of the text indicates the source of the strains; red represents food-derived Stx2l-STEC strains, and green represents human-derived Stx2l-STEC strains. The red stars indicate strains in this study. The insertion sites and downstream of prophages in the chromosomes are shown at the start and end of the prophages. The asterisk (*) signifies that downstream of the phage could not be conclusively identified.

3.4. SNP-Based Phylogenetic Relationship of Stx2l-Producing Strains

To assess the phylogenetic relationships of the Stx2l-STEC strains, whole-genome SNP-based phylogeny trees were constructed. The SNP analysis identified 1491 SNPs among the nine Stx2l-STEC strains. Two distinct clusters were observed based on the MLST types

(ST88 and ST23) of strains (Figure 3). Food-derived isolates from China and USA formed a cluster, raw mutton- and human-derived Stx2l-STEC strains from different countries were clustered. Pairwise SNP distance heatmaps were produced to illustrate the dissimilarity among strains. The SNP distance between the three food- and two human-derived strains in the ST23 cluster was ≤ 179, and it was ≤ 266 among four ST88 food-derived strains from China and USA.

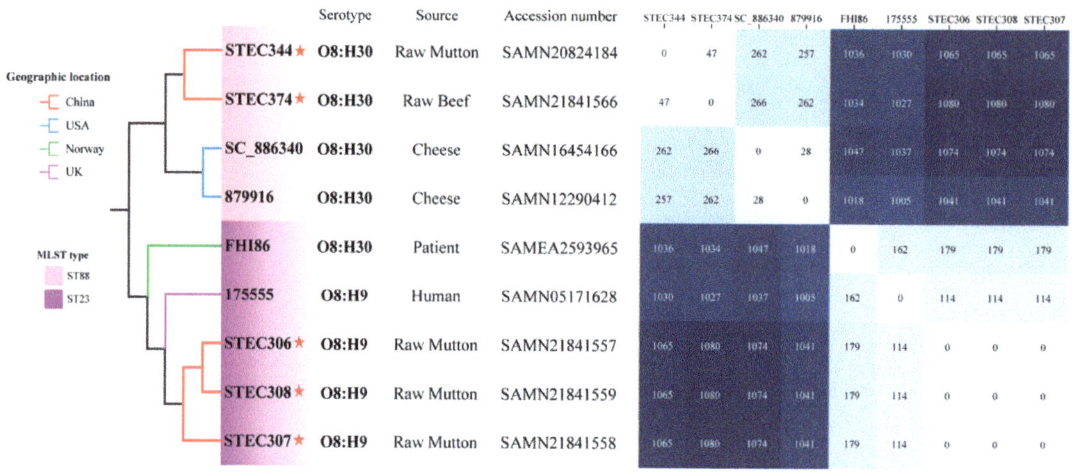

Figure 3. Phylogenetic tree based on single-nucleotide polymorphisms (SNPs) using the Maximum-Likelihood method. Branch length is ignored for better visualization. The red stars indicate strains in this study. The geographic location, MLST type, serotype, source, and accession number of all strains are shown. The heatmap illustrates pairwise SNP distances between the Stx2l-STEC strains.

4. Discussion

To our knowledge, this is the first study reporting the molecular traits of Stx2l-STEC strains. Different Stx subtypes vary in host specificity and toxicity, resulting in variations in pathogenic potentials to humans. Thus, precise Stx subtyping is valuable for risk assessment at an early stage after STEC infection. Since the standardization of Stx nomenclature and emergence of new Stx variants, novel Stx subtypes have been identified and a few Stx subtypes have been redesignated. The provisionally designed Stx2e variant Stx2e-O8-FHI-1106–1092 was redefined as Stx2l [7]. Therefore, we screened our STEC collection using the updated *stx*_subtyping database and identified five strains carrying the *stx2l* subtype, which was previously identified as *stx2e* based on the earlier nomenclature [7]. The sequence of transposable element IS2 was found among three of the five *stx2l* [17]. By searching the literature and publicly available sequences, several strains were found to carry the *stx2l* subtype, including the five meat-derived strains in this study, diarrheal patients-derived strains in Norway and Denmark [9], a human-derived strain in the UK, Roquefort cheese-derived strains in the USA [11], and sheep-derived strains in Ireland [15]. These data demonstrated a wide host range and geographical distribution of Stx2l-STEC strains.

We characterized the molecular characteristics of the Stx2l-STEC strains using WGS. The identical serogroup O8 was found among the five raw meat-derived strains in this study and all the publicly available genomes of Stx2l-STEC strains; notably, this serogroup was also possessed by the Stx2l reference strain FHI-1106-1092 and the patients-derived strains in Norway and Denmark [9], suggesting that O8 might be the dominant serogroup in the Stx2l-STEC strains. It should be noted that the O8:H30 Stx2l-STECs in this study showed a close phylogenetic relationship with the food-derived strains from USA; strains of this serotype have been isolated from diarrheal patients [9], suggesting the pathogenic potential of O8:H30 Stx2l-STEC strains. The whole-genome phylogeny showed that the patients-derived Stx2l-STEC strains from Norway and UK clustered with the three food-derived strains in this study; all these strains belonged to the MLST type ST23, indicating Stx2l-STECs of ST23 might be spread and, thus, possibly pose global public health risk. However, more Stx2l-STECs strains are needed to gain further insights. We further characterized the seven complete and two incomplete Stx2l-converting prophages; our data demonstrated a high similarity of the Stx2l prophages from different sources. It should be mentioned that the three mutton-sourced Stx2l-STECs shared nearly identical genomes; as these strains were isolated from different samples collected in the same sampling site, it is likely that they derived from the same clone.

In conclusion, our study demonstrated a potential wide distribution of Stx2l-STEC in diverse hosts and geographical regions. The genomic characterization revealed the genetic similarity of the Stx2l-STEC strains from different sources, with O8 possibly being the predominant serogroup. The genomes of the Stx2l-converting prophages from different sources were conserved. The Stx2l-STEC strains were phylogenetically clustered based on the sequence type of strains, and strains with ST23 might pose a public health risk.

Author Contributions: Conceptualization, Y.X. and X.B.; data curation, X.Y.; formal analysis, H.S.; funding acquisition, Y.X. and X.B.; methodology, X.Y.; project administration, Y.X. and X.B.; resources, H.S. and X.B.; software, X.Y. and Q.L.; supervision, Y.X. and X.B.; validation, X.Y.; visualization, Q.L.; writing—original draft, X.Y.; writing—review and editing, Y.X., A.M., and X.B. All authors have read and agreed to the published version of the manuscript.

Funding: This work was supported by the Ruth and Richard Julin's Foundation (Dnr: 2022-00277) and the National Natural Science Foundation of China (82072254).

Institutional Review Board Statement: The current study was reviewed and approved by the ethics committee of the National Institute for Communicable Diseases Control and Prevention, China CDC, with the number ICDC-2017003.

Informed Consent Statement: Not applicable.

Data Availability Statement: The genome data presented in this study are publicly available in NCBI with the accession numbers SAMN20824184, SAMN21841566, SAMN21841557, SAMN21841558, SAMN21841559, SAMN16454166, SAMN12290412, SAMEA2593965 and SAMN05171628.

Conflicts of Interest: The authors declare no conflict of interest.

References

1. Smith, J.L.; Fratamico, P.M.; Gunther, N.W.T. Shiga toxin-producing *Escherichia coli*. *Adv. Appl. Microbiol.* **2014**, *86*, 145–197. [PubMed]
2. Tarr, P.I.; Gordon, C.A.; Chandler, W.L. Shiga-toxin-producing *Escherichia coli* and haemolytic uraemic syndrome. *Lancet* **2005**, *365*, 1073–1086. [CrossRef]
3. Kruger, A.; Lucchesi, P.M. Shiga toxins and *stx* phages: Highly diverse entities. *Microbiology* **2015**, *161 Pt 3*, 451–462. [CrossRef] [PubMed]
4. Steyert, S.R.; Sahl, J.W.; Fraser, C.M.; Teel, L.D.; Scheutz, F.; Rasko, D.A. Comparative genomics and *stx* phage characterization of LEE-negative Shiga toxin-producing *Escherichia coli*. *Front. Cell Infect. Microbiol.* **2012**, *2*, 133. [CrossRef] [PubMed]
5. Rodríguez-Rubio, L.; Haarmann, N.; Schwidder, M.; Muniesa, M.; Schmidt, H. Bacteriophages of Shiga toxin-producing *Escherichia coli* and their contribution to pathogenicity. *Pathogens* **2021**, *10*, 404. [CrossRef]
6. Bryan, A.; Youngster, I.; McAdam, A.J. Shiga toxin producing *Escherichia coli*. *Clin. Lab. Med.* **2015**, *35*, 247–272. [CrossRef]
7. Scheutz, F.; Teel, L.D.; Beutin, L.; Piérard, D.; Buvens, G.; Karch, H.; Mellmann, A.; Caprioli, A.; Tozzoli, R.; Morabito, S.; et al. Multicenter evaluation of a sequence-based protocol for subtyping Shiga toxins and standardizing Stx nomenclature. *J. Clin. Microbiol.* **2012**, *50*, 2951–2963. [CrossRef]
8. Scheutz, F. Taxonomy Meets Public Health: The case of Shiga Toxin-Producing *Escherichia coli*. *Microbiol. Spectr.* **2015**, *2*, 15–36.
9. EFSA Biohaz Panel; Koutsoumanis, K.; Allende, A.; Alvarez-Ordóñez, A.; Bover-Cid, S.; Chemaly, M.; Davies, R.; De Cesare, A.; Herman, L.; Hilbert, F.; et al. Pathogenicity assessment of Shiga toxin-producing *Escherichia coli* (STEC) and the public health risk posed by contamination of food with STEC. *EFSA J.* **2020**, *18*, e05967.
10. Bai, X.; Fu, S.; Zhang, J.; Fan, R.; Xu, Y.; Sun, H.; He, X.; Xu, J.; Xiong, Y. Identification and pathogenomic analysis of an *Escherichia coli* strain producing a novel Shiga toxin 2 subtype. *Sci. Rep.* **2018**, *8*, 6756. [CrossRef]
11. Lacher, D.W.; Gangiredla, J.; Patel, I.; Elkins, C.A.; Feng, P.C. Use of the *Escherichia coli* identification microarray for characterizing the health risks of Shiga toxin-producing *Escherichia coli* isolated from foods. *J. Food Prot.* **2016**, *79*, 1656–1662. [CrossRef] [PubMed]

12. Gill, A.; Dussault, F.; McMahon, T.; Petronella, N.; Wang, X.; Cebelinski, E.; Scheutz, F.; Weedmark, K.; Blais, B.; Carrillo, C. Characterization of atypical Shiga toxin gene sequences and description of Stx2j, a new subtype. *J. Clin. Microbiol.* **2022**, *60*, e0222921. [CrossRef] [PubMed]
13. Yang, X.; Bai, X.; Zhang, J.; Sun, H.; Fu, S.; Fan, R.; He, X.; Scheutz, F.; Matussek, A.; Xiong, Y. *Escherichia coli* strains producing a novel Shiga toxin 2 subtype circulate in China. *Int. J. Med. Microbiol.* **2020**, *310*, 151377. [CrossRef] [PubMed]
14. Bai, X.; Scheutz, F.; Dahlgren, H.M.; Hedenström, I.; Jernberg, C. Characterization of clinical *Escherichia coli* strains producing a novel Shiga toxin 2 ubtype in Sweden and Denmark. *Microorganisms* **2021**, *9*, 2374. [CrossRef] [PubMed]
15. McCarthy, S.C.; Macori, G.; Duggan, G.; Burgess, C.M.; Fanning, S.; Duffy, G. Prevalence and whole-genome sequence-based analysis of Shiga toxin-producing *Escherichia coli* isolates from the Recto-Anal junction of slaughter-age Irish sheep. *Appl. Environ. Microbiol.* **2021**, *87*, e0138421. [CrossRef]
16. Bai, X.; Wang, H.; Xin, Y.; Wei, R.; Tang, X.; Zhao, A.; Sun, H.; Zhang, W.; Wang, Y.; Xu, Y.; et al. Prevalence and characteristics of Shiga toxin-producing *Escherichia coli* isolated from retail raw meats in China. *Int. J. Food Microbiol.* **2015**, *200*, 31–38. [CrossRef]
17. Yang, X.; Wu, Y.; Liu, Q.; Sun, H.; Luo, M.; Xiong, Y.; Matussek, A.; Hu, B.; Bai, X. Genomic characteristics of Stx2e-producing *Escherichia coli* strains derived from humans, animals, and meats. *Pathogens* **2021**, *10*, 1551. [CrossRef]
18. Aziz, R.K.; Bartels, D.; Best, A.A.; DeJongh, M.; Disz, T.; Edwards, R.A.; Formsma, K.; Gerdes, S.; Glass, E.M.; Kubal, M.; et al. The RAST Server: Rapid annotations using subsystems technology. *BMC Genom.* **2008**, *9*, 75. [CrossRef]
19. Darling, A.E.; Mau, B.; Perna, N.T. progressiveMauve: Multiple genome alignment with gene gain, loss and rearrangement. *PLoS ONE* **2010**, *5*, e11147. [CrossRef]
20. Sullivan, M.J.; Petty, N.K.; Beatson, S.A. Easyfig: A genome comparison visualizer. *Bioinformatics* **2011**, *27*, 1009–1010. [CrossRef]
21. Croucher, N.J.; Page, A.J.; Connor, T.R.; Delaney, A.J.; Keane, J.A.; Bentley, S.D.; Parkhill, J.; Harris, S.R. Rapid phylogenetic analysis of large samples of recombinant bacterial whole genome sequences using Gubbins. *Nucleic Acids Res.* **2015**, *43*, e15. [CrossRef] [PubMed]
22. Pinto, G.; Sampaio, M.; Dias, O.; Almeida, C.; Azeredo, J.; Oliveira, H. Insights into the genome architecture and evolution of Shiga toxin encoding bacteriophages of *Escherichia coli*. *BMC Genom.* **2021**, *22*, 366. [CrossRef] [PubMed]

microorganisms

MDPI

Article

Antimicrobial Resistance, Enterotoxin and *mec* Gene Profiles of *Staphylococcus aureus* Associated with Beef-Based Protein Sources from KwaZulu-Natal Province, South Africa

Thembeka Thwala [1], Evelyn Madoroba [1], Tsolanku S. Maliehe [1], Kudakwashe Magwedere [2], Albert K. Basson [1] and Patrick Butaye [3,4,*]

1 Department of Biochemistry and Microbiology, University of Zululand, Private Bag X1001, KwaDlangezwa, Empangeni 3886, South Africa; thembekaprincess6@gmail.com (T.T.); madorobae@unizulu.ac.za (E.M.); maliehet@unizulu.ac.za (T.S.M.); bassona@unizulu.ac.za (A.K.B.)

2 Directorate of Veterinary Public Health, Department of Agriculture, Land Reform and Rural Development, Pretoria 0001, South Africa; kudakwashem@dalrrd.gov.za

3 Department of Pathobiology, Pharmacology and Zoological Medicine, Faculty of Veterinary Medicine, Ghent University, Salisburylaan 133, B9820 Merelbeke, Belgium

4 Department of Biomedical Sciences, Ross University School of Veterinary Medicine, Basseterre P.O. Box 334, Saint Kitts and Nevis

* Correspondence: patrick.butaye@ugent.be

Abstract: Annually, approximately 23,000 cases of food poisoning by *Staphylococcus aureus* enterotoxins are reported worldwide. The aim of this study was to determine the occurrence and characterize *S. aureus* on beef and beef products in South Africa. Organ meats (n = 169), raw processed meat (n = 110), raw intact (n = 53), and ready-to-eat meats (n = 68) were obtained from 25 retail outlets. *S. aureus* was isolated and enumerated according to the ISO 6888-1 method. Identification of the strains was performed by MALDI-TOF MS. The antimicrobial resistance was determined using the disc diffusion test. The presence of methicillin-resistance genes and the staphylococcal enterotoxin genes was determined by PCR. Prevalence was low (13/400; CI 1.7–5) and all but one positive sample were from organ meats. Eight isolates were resistant to at least one antibiotic. Two isolates carried the *mecC* gene. All the isolates tested positive for *seg*, *seh*, *sei*, and *sep*, whilst 53.8% were positive for *sea*. None of the isolates was positive for *ser*, *sej*, *seb*, *sec*, or *sed*. The prevalence of *S. aureus* was low, with organ meats being the most contaminated. The presence of *mecC*-positive MRSA and of enterotoxins warrants further investigation and risk assessment.

Keywords: methicillin-resistant *Staphylococcus aureus*; staphylococcal enterotoxin genes; antimicrobial resistance; beef-based products; *S. aureus* contamination; food safety

Citation: Thwala, T.; Madoroba, E.; Maliehe, T.S.; Magwedere, K.; Basson, A.K.; Butaye, P. Antimicrobial Resistance, Enterotoxin and *mec* Gene Profiles of *Staphylococcus aureus* Associated with Beef-Based Protein Sources from KwaZulu-Natal Province, South Africa. *Microorganisms* **2022**, *10*, 1211. https://doi.org/10.3390/microorganisms10061211

Academic Editor: Agapi I. Doulgeraki

Received: 9 May 2022

Accepted: 10 June 2022

Published: 14 June 2022

Publisher's Note: MDPI stays neutral with regard to jurisdictional claims in published maps and institutional affiliations.

1. Introduction

Beef is known for its role in supplying protein, minerals, and vitamins in human nutrition [1]. Due to its high nutritional content, beef is an excellent substrate for the growth of microorganisms, of which some are leading causes of meat spoilage [2]. Spoilage of meat is enhanced by inadequately stored or packed meat [3]. Different storage conditions, such as cold and gaseous composition, on packed meat are most likely to suppress the microflora, among them *S. aureus* [4].

S. aureus causes staphylococcal food poisoning (SFP) through the ingestion of food contaminated with staphylococcal enterotoxins [5]. This enterotoxaemia is characterized by diarrhea, nausea, abdominal cramping, and vomiting within 24 h of eating [6]. Contamination of food by *S. aureus* may originate from the animal, the food handlers, and the environment. It may be a consequence of poor hygiene during processing from slaughter to final product or inappropriate storage and household manipulations; however, contamination of meat is a complicated process which may occur well before the meat reaches retail outlets [7]. In addition to toxins encoded by the *seb*, *sec*, *sed*, and *see* genes, in particular, strains that produce the Staphylococcal Enterotoxin A (SEA), encoded by the *sea* gene, have caused a large number of outbreaks [8,9].

Antimicrobial resistance (AMR) is an increasing global challenge mainly driven by the overall use of antimicrobials [10]. In certain *S. aureus* clones, AMR is a major problem, especially in methicillin-resistant *S. aureus* (MRSA), of which the prevalence increases globally [11]. In the last 15 years, MRSA clonal complex 398 was discovered in food-producing animals, while other sequence types (ST), such as ST1, ST5, ST9, ST97, ST130, and ST433, have been reported to a lesser extent [12]. These strains were subsequently named Livestock Associated (LA)-MRSA. LA-MRSA Clonal complex 398 (CC 398) is mainly prevalent in Europe and North America; however, it has been reported in Asia as well as in Africa [12]. To a lesser extent, LA-MRSA CC398 has been associated with different infections in humans, including skin and soft tissue illnesses, ventilator-associated pneumonia, and septicemia [13].

There are few studies that have been conducted on meat and meat products in Africa so far [14,15]; most studies did not type the isolates, and only few of the studies have identified AMR genes [14–17]. Therefore, this study aimed to determine the occurrence, AMR, and virulence genes of *S. aureus* isolated from beef and beef products in retail outlets of the KwaZulu-Natal province, South Africa.

2. Materials and Methods

2.1. Ethical Approval

Ethical approval for this study was obtained from the University of Zululand with certificate number UZREC 171110-030 PMG 2019/112.

2.2. Study Design

The cross-sectional study involved the collection and microbiological analysis of meat and meat products from retail outlets and butcheries from King Cetshwayo and iLembe districts in the KwaZulu-Natal province of South Africa. King Cetshwayo district covers an area of 8213 km^2 from the north coast region, whereas iLembe covers an area of 3269 km^2 from the south coast region with a population of approximately 885,944 and 606,809, respectively [18]. The two districts contribute to about a quarter of the total population in KwaZulu-Natal KZN (Figures 1 and 2).

Figure 1. Geographical map representing King Cetshwayo district, KwaZulu-Natal. Source: https://municipalities.co.za/map/124/king-cetshwayo-district-municipality (accessed on 1 November 2020) [19].

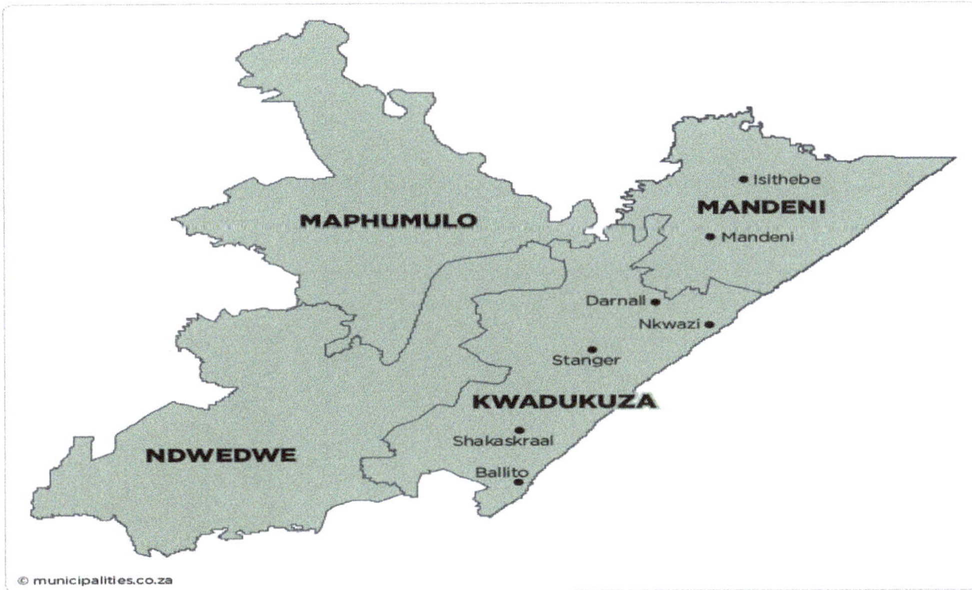

Figure 2. Geographical map representing iLembe district, KwaZulu-Natal. Source: https://municipalities.co.za/map/117/ilembe-district-municipality (accessed on 1 November 2020) [19].

2.3. Sample Size Determination

There are important statistical variables to consider when determining the sample size for a surveillance study [20]. These include z, α, p, and d, where z (1.96) is the normal deviate for two-tailed alternative hypotheses [21]. Alpha (α) is the level of significance and it is usually 5%, which implies that having a 5% probability of incorrectly rejecting a null hypothesis is acceptable [22]. The *p*-value is the expected prevalence proportion, and a prevalence of 50% (0.50) was assumed in this study based on a national surveillance of foodborne pathogens in South Africa by [23], which detected about 56% *S. aureus* in diverse meat and meat products from various establishments. The value of d is precision, and at the confidence interval of 95%, d is 0.05. In this study, the following formula was used to calculate the sample size of the surveillance study:

$$\text{Sample size } = \frac{z\left(1-\frac{\alpha}{2}\right)^2 p(1-p)}{d^2} = \frac{1.96^2\ 0.50(1-0.50)}{0.05^2} = 384 \quad (1)$$

However, 400 samples were collected in this study for robust results.

2.4. Sample Collection

A total of 400 samples were collected during the cross-sectional study. Twenty-five retail outlets and butcheries from King Cetshwayo and iLembe districts were included in this study. The beef samples were ready-to-eat beef products (n = 68), raw processed beef (n = 110), raw intact beef (n = 53), and organ meats (n = 169). Samples were packed into sampling bags using strict aseptic techniques, and placed in cooler bags containing ice packs to maintain a temperature of approximately 4 °C. The packaged samples were transported immediately to the Microbiology laboratory at the University of Zululand, Department of Biochemistry and Microbiology, for further bacteriological examination. Samples were analyzed immediately after arrival.

2.4.1. Microbiological Analysis

Control Strains for Quality Control

The *S. aureus* ATCC 25,923 (Microbiologics, MN, USA) and field strains (positive for tested virulence factors) were included in all laboratory experiments as positive control strains. ATCC 25,922 was used as negative control.

Detection, Enumeration, and Isolation and Identification of *S. aureus*

The detection, enumeration, and isolation of *S. aureus* was performed according to the ISO 6888-1:1999 AMD 2018 standard method [24]. Briefly, each sample was analyzed for the presence of *S. aureus* by weighing 25 g, followed by addition of 225 mL of buffered peptone water. The samples and buffered peptone water were thoroughly mixed in a homogenizer (Bagmixer 400 cc, Interscience, France) for 2 min at 10 stroke/s. Subsequently, ten-fold serial dilutions were made using sterile pipettes [25]. From these dilutions, 0.1 mL was inoculated in duplicate onto Baird Parker agar plates (Oxoid, UK) using the spread plate technique, as described by Goja et al. [25]. Plates were incubated at 37 °C for 24 to 48 h After incubation, the typical colonies were counted. Typical *Staphylococcus* spp. appeared as shiny black colonies [26]. To calculate the number of colony-forming units per gram (CFU/g), the colonies on the countable plate were multiplied by final dilution factor. The presumptive colonies were purified three times through sub-culturing on nutrient agar (Oxoid, UK) and incubation at 37 °C for 24 h. The presumptive *S. aureus* colonies were subjected to Gram staining, catalase test, mannitol salt, and free and bound coagulase tests (Oxoid, UK) [27]. Gram-positive cocci that appeared purple with grape-like shape were catalase-positive, appearing yellow on mannitol salt agar due to mannitol fermentation, and where coagulase-positive, were considered to be presumptive *S. aureus* and the colonies were subjected to further tests. Identification of *S. aureus* was confirmed using MALDI-TOF MS, according to the manufactures instructions for the MALDI Biotyper®(Bruker Daltonics, Germany) [28]. All confirmed *S. aureus* isolates were streaked on 5% sheep blood agar plates to identify the type of hemolysin they produce [29]. The plates were incubated at 37 °C for 24 h ± 2 [30].

2.5. Antimicrobial Susceptibility Testing

For the antimicrobial susceptibility test, the Kirby Bauer disk diffusion method according to Clinical Laboratory Standard Institute guidelines was applied [31,32]. Briefly, from a pure bacterial culture, 2–5 colonies were suspended in 5 mL sterile saline solution The bacterial concentration was adjusted to an optical density of 0.5 on the McFarland scale [31,33]. A sterile cotton swab was dipped into the suspension and excess fluid was removed by squeezing the swab at the top of the bijou bottles. The bacteria were inoculated onto Mueller Hinton agar (Thermofisher, Waltham, MA, USA) by streaking in three different directions to obtain confluent bacterial growth. The medium surface was allowed to dry, followed by placing the following antimicrobial disks: ciprofloxacin (5 µg), cefoxitin (30 µg), clindamycin (2 µg), erythromycin (15 µg), rifampicin (5 µg), oxacillin (1 µg), kanamycin (30 µg), penicillin G (10 units), chloramphenicol (30 µg), gentamicin (10 µg), and trimethoprim (25 µg) (Davies Diagnostics, Randburg, South Africa) [34].

2.6. Detection of Selected Resistance and Virulence Genes

2.6.1. DNA Extraction and PCR for Staphylococcal Enterotoxins and *mec* Genes

The Zymo DNA extraction kit (California, CA, USA) was used for the extraction of the DNA according to the manufacturer's instructions. The quality and quantity of the DNA were measured using a Nanodrop 2000 spectrophotometer (Thermofisher Scientific, Waltham, MA, USA).

The enterotoxin genes (*sea, seb, sed, sec, she, seg, ser, sei, sep, sej*) and *mec* genes (*mecA* and *mecC*) were assessed by PCR using the primers listed in Table 1. The 20 µL PCR reaction mixtures contained 10–30 ng of template DNA (in 1 µL), NEB one Taq 2× master mix with standard buffer (10 µL), forward primer (1 µL), reverse primer (1 µL), and nuclease free water (7 μL). PCR amplifications for *sea, seb, sec, sed,* and *ser* were carried out in a thermal cycler with the following thermal conditions: initial denaturation for 5 min at 95 °C; 35 cycles of 30 s at 94 °C, 40 s at 56 °C, and 1 min at 68 °C; final extension for 5 min at 68 °C. The PCR conditions for *seg, sei, sep, sej* and *she* were similar to the above, except that the annealing stage was performed at 53 °C for 40 s.

Table 1. Oligonucleotide sequence primers used to target genes for species confirmation in *S. aureus*.

Target Gene	Primer Sequence (5′-3′)	Product Size (bp)	Reference
tsst	5′ CATCTACAAACGATAATATAAAGG 3′ CATTGTTATTTTCCAA TAACCACCC	481	[35]
mecA	5′ GAA ATG ACT GAA CGT CCG AT 3′ CTG GAA CTT GTT GAG CAG AG	399	[35]
sea	5′ GGTTATCAATGTGCGGGTGG 3′ CGGCACTTTTTTCTCTTCGG	102	[36]
seb	5′ GTATGGTGGTGTAACTGAGC 3′ CCAAATAGTGACGAGTTAGG	164	[36]
sec	5′ AGATGAAGTAGTTGATGTGTATGG 3′ CACACTTTTAGAATCAACCG	451	[36]
sed	5′ CCAATAATAGGAGAAAATAAAA 3′ ATTGGTATTTTTTTTCGTTC	278	[36]
ser	5′ AGATGTGTTTGGAATACCCTAT 3′ CTATCAGCTGTGGAGTGCAT	123	[37]
seg	5′ GTTAGAGGAGGTTTTATG 3′ TTCCTTCAACAGGTGGAGA	198	[37]
she	5′ CAACTGCTGATTTAGCTCAG 3′ CCCAAACATTAGCACCA	173	[38]
sei	5′ GGCCACTTTATCAGGACA 3′ AACTTACAGGCAGTCCA	328	[37]
sej	5′ GTTCTGGTGGTAAACCA 3′ GCGGAACAACAGTTCTGA	131	[37]
sep	5′ TCAAAAGACACCGCCAA 3′ ATTGTCCTTGAGCACCA	396	[39]
mecC	5′ GAAAAAAAGGCTTAGAACGCCTC 3′ GAAGATCTTTTCCGTTTTCAGC	138	[17]

2.6.2. Agarose Gel Electrophoresis

PCR products were subjected to electrophoresis in 1.5% agarose gels stained with ethidium bromide at 3 volts/cm for approximately 30 min. The 50 bp and 100 bp DNA ladders were used to estimate the size of PCR amplicons. The PCR amplicons were visualized under ultraviolet light and the gel images were documented using a gel documentation system (E-Box).

3. Results

3.1. Prevalence of S. Aureus in Meat

Out of the 400 beef and beef products that were analyzed, 3.25% (n = 13; CI 1.7–5) tested positive for *S. aureus* (Table 2). From each of the 13 positive samples, one isolate was retained for further investigation. The *S. aureus*-positive samples were predominantly organ meats (n = 10/13; CI 46.2–95), followed by raw intact beef (n = 2/13; 1.9–45). Only one of the 13 *S. aureus*-positive samples was from ready-to-eat beef. No *S. aureus* was detected in raw processed beef. The 13 *S. aureus* from 13 positive samples showed alpha, beta, and gamma hemolysis reactions on 5% sheep blood agar.

3.2. Enumeration of Staphylococcus aureus

Table 3 shows the results of *Staphylococcus aureus* enumeration of the 13 positive samples. The *S. aureus* counts from beef-based products ranged from 2.65 $\log_{10}$ CFU/g to

4.1 log_{10} CFU/g (Table 4). *S. aureus* from 1 of the 13 positive samples (ox kidneys) were too numerous to count.

Table 2. Prevalence of *S. aureus* from beef-based products in selected districts from KZN.

Meat Type	Number of Samples	Number Positive	Prevalence (%)	CI *
Organ meat	169	10	5.9	2.9–11
Raw intact meat	53	2	3.8	0.5–13
Processed meat	110	0	0	0–3
Ready to-eat meat	68	1	1.5	0–8
Total	400	13	3.25	1.7–5

CI * refers to confidence intervals.

Table 3. *S. aureus* counts from beef and beef products.

Sample Number	Sample ID	Sample Type	Log_{10} CFU/g
176	KkwaAO176	Ox kidneys	3.82
167	KGinEO167	Ox tripe (bible)	4.07
174	KkwaAO174	Ox liver	3.65
200	KkwaLO200	Ox lungs	3.34
177	KkwaAO177	Ox kidneys	3.87
201	KkwaLO201	Ox lungs	3.62
235	KmelEI235	Beef steak tender	4.03
250	KmelDO250	Ox lungs	2.65
370	ILestanBR370	Biltong	3.37
98	KRbyEI98	Stewing beef	4.1
162	KGinEO162	Ox liver	3.38
238	KmelEO238	Ox liver	3.23
302	ILemanEO302	Ox kidneys	uncountable

Table 4. Antimicrobial resistance among 13 *S. aureus* (including MRSA) from meat and meat products in selected KZN province municipalities.

Antimicrobial Classes	Antimicrobial Agents	Number of Tested Isolates	Number of Resistant Isolates	Percentage % (CI)
Penicillins	Penicillin G	13	5	38.46 (13.9–68)
Penicillin-resistant penicillins	Oxacillin	13	1	7.69 (0.2–36)
Cephalosporins	Cefoxitin	13	1	7.69 (0.2–36)
Aminoglycosides	Gentamicin	13	0	0 (0–25)
	Kanamicin	13	0	0 (0–25)
Macrolides	Erythromicin	13	3	23.08 (5–54)
Lincosamides	Clindamicin	13	4	30.77 (9.1–61)
Fluoroquinolones	Ciprofloxacin	13	2	15.38 (1.9–45)
Phenicols	Chloramphenicol	13	0	0 (0–25)
Folate-pathway inhibitor	Trimethoprim	13	0	0 (0–25)
Rifampin	Rifampicin	13	1	7.69 (0.2–36)
Tetracyclines	Tetracycline	13	1	7.69 (0.2–36)

3.3. *Antimicrobial Susceptibility Testing*

AMR of the *S. aureus* isolates are shown in Table 4. Eight out of 13 (61.54%) isolates were resistant to at least one antibiotic. Less than 50% of the isolates exhibited resistance to penicillin G (38.46%; $n = 5/13$; CI 13.9–68), cefoxitin (7.69%; $n = 1/13$; CI 0.2–36), tetracycline (7.69%; $n = 1/13$; 0.2–36), oxacillin (7.69%; $n = 1/13$; CI 0.2–36), clindamycin (30.76%; $n = 4/13$; CI 9.1–61), erythromycin (23.07%; $n = 3/13$; CI 5–54), ciprofloxacin (15.38%; $n = 2/13$; CI 1.9–45), and rifampicin with a resistance percentage of 7.6%. ($n = 1/13$; CI 0.2–36). Multi-drug resistance (MDR), which is the lack of susceptibility to at least three antimicrobial classes [40], was observed in two *S. aureus* isolates. Two MDR profiles were observed, namely, PG-FOX-OX-RP-CD-E-TET ($n = 1$) and PG-E-CD ($n = 1$).

3.4. Detection of Selected Resistance and Virulence Genes

3.4.1. Methicillin-Resistant Determinants

All isolates were tested for the presence of *mecA* and *mecC* genes. None of the isolates were positive for *mecA* genes. Two isolates (15.4%; CI 1.9–45) tested positive for *mecC* gene (Figure 3).

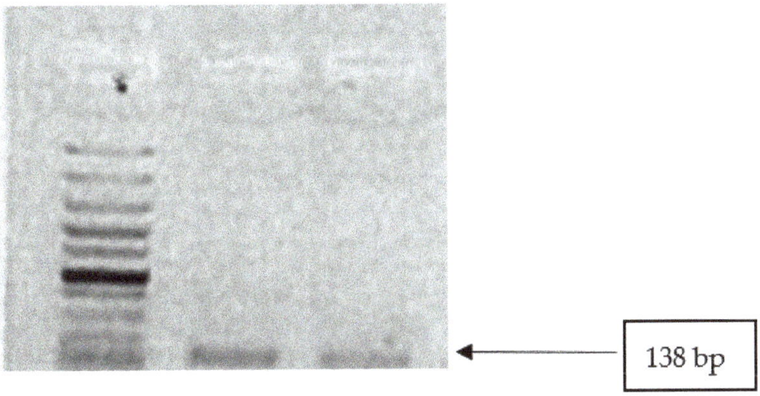

Figure 3. Image showing *mecC* gene amplicons observed on agarose gel. Lane 1: 100 bp DNA ladder; lanes 2–3 show positive band for *mecC* genes (138 bp).

3.4.2. *S. aureus* Enterotoxin Genes

In Table 5, Figures 4 and 5, the results of the virulence genes are shown. Out of eleven enterotoxin genes that were tested, five (*sep, seh, sei, sej, sea*) were detected. All *S. aureus* strains were positive for *seg, seh,* and *sei, sep*. The *sea* gene was detected in 7 of the 13 *S. aureus* (53.84%).

Table 5. PCR amplification results for *S. aureus* methicillin-resistance and enterotoxin genes.

Sample Code	Sample	***S. aureus* Enterotoxin, *mec* Genes and Resistance Profile**														
		sec	*sed*	*seg*	*sep*	*sej*	*seh*	*sei*	*ser*	*sea*	*tsst*	*seb*	*sed*	*mecC*	*mecA*	**R-Profile**
KRbyEI98	Stewing beef	-	-	+	+	-	+	+	-	+	-	-	-	-	-	N
KGinEO162	Liver	-	-	+	+	-	+	+	-	+	-	-	-	+	-	N
KGinEO167	Tripe (omasum)	-	-	+	+	-	+	+	-	+	-	-	-	-	-	PG-CD-E
KkwaAO174	Liver	-	-	+	+	-	+	+	-	+	-	-	-	-	-	CIP
KkwaAO176	Kidneys	-	-	+	+	-	+	+	-	+	-	-	-	+	-	PG-FOX-TET-OX-CD-E-RP
KkwaAO177	Kidneys	-	-	+	+	-	+	+	-	+	-	-	-	-	-	PG-E
KkwaLO200	Lungs	-	-	+	+	-	+	+	-	-	-	-	-	-	-	N
KkwaLO201	Lungs	-	-	+	+	-	+	+	-	-	-	-	-	-	-	N
KmelEI235	Beef steak	-	-	+	+	-	+	+	-	-	-	-	-	-	-	CD
KmelEO238	Liver	-	-	+	+	-	+	+	-	-	-	-	-	-	-	PG
KmelDO250	Lungs	-	-	+	+	-	+	+	-	-	-	-	-	-	-	CD
ILemanEO302	Kidneys	-	-	+	+	-	+	+	-	-	-	-	-	-	-	PG-CIP
ILestanBR370	Biltong	-	-	+	+	-	+	+	-	+	-	-	-	-	-	N

N: susceptible to all antibiotics; R-profile: antimicrobial resistance profile for each isolate; FOX (cefoxitin); OX (oxacillin); PG (penicillin G); CIP (ciprofloxacin); TET (tetracycline); E (erythromycin); RP (rifampicin); CD (clindamycin).

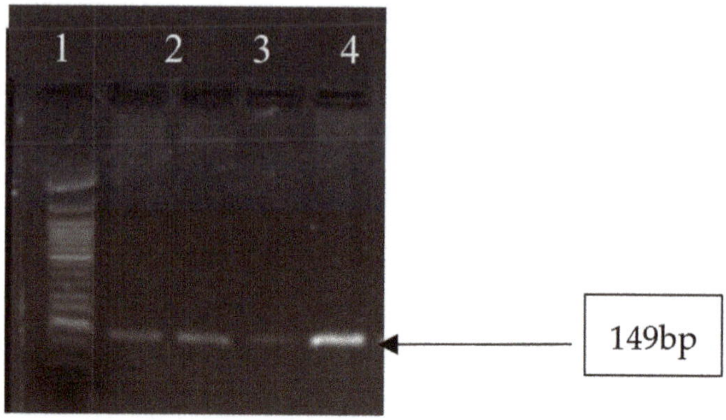

Figure 4. Image showing *seg* gene amplicons observed on agarose gel. Lane 1: 50 bp DNA ladder lanes 2–5 show amplicon sizes for samples that were positive for *seg* gene (149 bp).

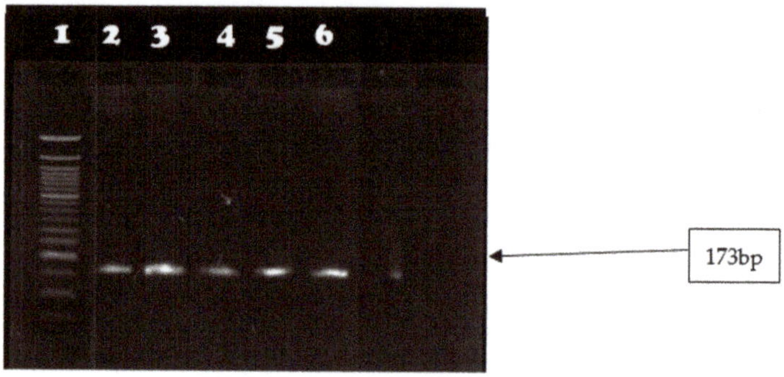

Figure 5. Image showing *seh* gene amplicons observed on agarose gel. Lane 1: 50 bp DNA ladder, lanes 2–6 show amplicon sizes for samples that were positive for *seh* gene (165 bp).

4. Discussion

The aim of this study was to determine the occurrence, AMR, and virulence characteristics of *S. aureus* from products of bovine origin in retail outlets of selected municipalities. Though most studies in Africa found prevalences of between 15 and 40%, with the exception of 55% *S. aureus* detection in Algeria, the overall occurrence of 3.25% from this study is lower compared to most of the previous African studies [41–45]. Higher prevalences of up to 65% have also been found in non-African countries such as Turkey, Jordan, United States of America (USA), and several countries in Europe [46–52]. It is important to note that differences in methodology and sample size should be taken into account and may explain, in part, the differences seen [53,54]. It is important to apply a system such as the ISO standard, used in this study, to allow direct comparison of the prevalence between different studies.

The relatively high *S. aureus* prevalence in organ meats (kidneys, livers, lungs) compared to raw intact beef meat and ready to-eat meat was conspicuous in this study (though not to a significant extent). Probably the organ meat may be more prone to cross contamination compared to other meat types, and *S. aureus* can be found in the intestines [55]. *S. aureus* occurrence was also observed in ready to-eat meats and this may be attributed to cross contamination and growth due to further preparation [56,57]. The differences in preparation and the type of preparation of the ready-to-eat (RTE) beef, as well as conservation of the product, may play a large influence.

The average counts of the *S. aureus*-positive samples in this study ranged from 2.65 $\log_{10}$ and 4.07 $\log_{10}$ per gram in organ meat. These *S. aureus* counts are lower than those that were previously observed for organ beef meat in South Africa (5.1 log–log 5.6) [14].

When considering the *S. aureus* limit of 100 CFU/g in RTE, proposed by the guidelines for environmental health officers on the interpretation of microbiological analysis data of food [58], the 13 positive samples were not within the compliance limits, though they were sold at retail level. The situation is concerning for RTE biltong, which is not processed further prior to consumption. The contaminated samples were mainly plucked meats that might not be subjected to similar strict hygiene scrutiny as beef cuts. It is possible that the *S. aureus* counts may have increased in the pluck meats during transportation, probably due to an inadequate cold chain, or the level of preservation at the retail level may have contributed to an increase in bacterial numbers.

As we found only few strains, comparing with other studies is difficult. The isolates from this study were, in general, more susceptible to antimicrobials than those from other studies on beef in South Africa [15] and other African countries [42,44], but are similar to studies in Europe [50], and higher than what has been detected in the United States of America [59].

Interestingly, two isolates were MRSA, with only one detected phenotypically. Phenotypic methicillin resistance should always be confirmed by PCR, as false positive and false negative results may be obtained by the phenotypic tests. While this may not seem a lot, it may have a significant public health impact. The MRSA isolates from this study were *mecC*-positive. While this resistance gene has not been associated extensively with MRSA either in humans or animals, it has, however, been isolated mainly from animals, including wildlife [60]. In most countries, the *mecA* gene is mostly found in MRSA [15,61–65]. However, in South Africa, the *mecC* gene has also been shown as the sole methicillin-resistance gene in strains from different food-producing animal species as well as wild birds [17]. This might indicate a very specific and unique situation in South Africa and urges for a more large-scale study of MRSA on food-producing animals, wild animals, and foods derived from animals in South Africa so as to determine the human health hazard. These studies should include whole-genome sequencing to determine their true epidemiology.

Staphylococcal enterotoxins (SEs) types SEA to SEE have been reported to account for approximately 95% of food poisoning outbreaks caused by staphylococci [66], whilst the remainder may be due to the other SE types, including SEG, SEH, SEI, SEJ, SEK, SEL, SEM, SEN, and SEO [5]. Based on the positive enterotoxin genes, it is clear that many *S. aureus* isolates from this study are enterotoxigenic. Some of the genes found in this study, such as *sea*, *seh*, *seg*, and *sei*, have been associated with outbreaks of food poisoning in different parts of the world [5,67–70]. However, *seg* and *sei* have not frequently been isolated from food isolates and are, rather, associated with staphylococcal scarlet fever and toxic shock syndrome [71]. The *seg*, *sei*, and *seh* have also been identified in patients with other *S. aureus*-associated infections [72].

In the current study, all the 13 *S. aureus* isolates tested positive for *seg*, *sei*, and *seh* genes. The *seg* and *sei* genes are components of the *egc* operon, together with *sem*, *sen*, and *seo* enterotoxin genes [70]. The *egc* operon is located on a mobile genetic element (MGE) [70] and can thus be transferred to non-pathogenic *S. aureus* [60]. This combination is, however, rarely found in strains involved in toxi-infections [5,38,46,71].

5. Conclusions

In conclusion, the current study contributes to the knowledge about *S. aureus* on beef in South African markets. While the overall prevalence was relatively low, care should, however, be taken when handling pluck meats to avoid cross contamination with utensils, working surfaces, and RTE. Few *S. aureus* isolates exhibited antimicrobial resistance; however, the presence of *mecC*-positive *S. aureus* strains is worrisome. Five classical staphylococcal enterotoxin genes were identified from these isolates, which indicate a health risk to the consumers. The observation of *mecC*-positive MRSA that are present on food and have been reported also in food-producing animals warrants a One Health study on MRSA in food-producing animals, pet animals, wildlife, and foods in South Africa. These studies should include whole-genome sequencing so as to determine the epidemiology and origins of *mecC*-positive MRSA.

Author Contributions: Conceptualization, E.M. and P.B.; methodology, T.T., E.M., P.B. and T.S.M.; validation, E.M. and P.B.; formal analysis, T.T., E.M. and P.B.; investigation, T.T.; resources, E.M.; data curation, E.M.; writing—original draft preparation, T.T.; writing—review and editing, E.M., P.B., A.K.B. and K.M.; supervision, E.M., A.K.B. and P.B.; project administration, E.M.; funding acquisition, E.M. All authors have read and agreed to the published version of the manuscript.

Funding: This research was funded by Red Meat Research and Development South Africa (RMRD SA), and the Department of Trade Industry and Competition grant number THRIP/22/30/11/2017.

Data Availability Statement: Not applicable.

Acknowledgments: We thank the relevant municipalities that were involved in the study.

Conflicts of Interest: The authors declare no conflict of interest.

References

1. Biesalski, H.K. Meat as a component of a healthy diet-are there any risks or benefits if meat is avoided in the diet? *Meat Sci.* **2005**, *70*, 509–524. [CrossRef] [PubMed]
2. Barros, M.A.F.; Nero, L.A.; Monteiro, A.A.; Beloti, V. Technology, Identification of main contamination points by hygiene indicator microorganisms in beef processing plants. *J. Food Sci. Technol.* **2007**, *27*, 856–862. [CrossRef]
3. Koutsoumanis, K.; Sofos, J.N. Microbial Contamination of carcasses and cuts. In *Encyclopedia of Meat Sciences*; Jensen, W.K., Devine, C., Dikeman, M., Eds.; Elsevier Academic Press: Amsterdam, The Netherlands, 2004; pp. 727–737.
4. Moutiq, R.; Misra, N.; Mendonca, A.; Keener, K. In-package decontamination of chicken breast using cold plasma technology: Microbial, quality and storage studies. *Meat Sci.* **2020**, *159*, 107942. [CrossRef]
5. Omoe, K.; Ishikawa, M.; Shimoda, Y.; Hu, D.L.; Ueda, S.; Shinagawa, K. Detection of *seg, seh,* and *sei* genes in *Staphylococcus aureus* isolates and determination of the enterotoxin productivities of *S. aureus* isolates harboring *seg, seh,* or *sei* genes. *J. Clin. Microbiol.* **2002**, *40*, 857–862. [CrossRef] [PubMed]
6. Abdalrahman, L.S.; Fakhr, M.K. Incidence, antimicrobial susceptibility, and toxin genes possession screening of *Staphylococcus aureus* in retail chicken livers and gizzards. *Foods* **2015**, *4*, 115–129. [CrossRef]
7. Madoroba, E.; Gelaw, A.K.; Kapeta, D. *Salmonella* contamination, serovars and antimicrobial resistance profiles of cattle slaughtered in South Africa. *J. Vet. Res.* **2016**, *83*, 1–8. [CrossRef]
8. Hennekinne, J.A.; De Buyser, M.L.; Dragacci, S. *Staphylococcus aureus* and its food poisoning toxins: Characterization and outbreak investigation. *FEMS Microbiol. Rev.* **2012**, *36*, 815–836. [CrossRef]
9. Omwenga, I.; Aboge, G.O.; Mitema, E.S.; Obiero, G.; Ngaywa, C.; Ngwili, N.; Wamwere, G.; Wainaina, M.; Bett, B. *Staphylococcus aureus* enterotoxin genes detected in milk from various livestock species in northern pastoral region of Kenya. *Food Control* **2019**, *103*, 126–132. [CrossRef]
10. Roca, I.; Akova, M.; Baquero, F.; Carlet, J.; Cavaleri, M.; Coenen, S.; Cohen, J.; Findlay, D.; Gyssens, I.; Heuer, O.E.; et al. The global threat of antimicrobial resistance: Science for intervention. *New Microbes New Infect.* **2015**, *6*, 22–29. [CrossRef]
11. Stefani, S.; Goglio, A. Methicillin-resistant *Staphylococcus aureus*: Related infections and antibiotic resistance. *Int. J. Infect. Dis.* **2010**, *14* (Suppl. S4), S19–S22. [CrossRef]
12. Butaye, P.; Argudín, M.A.; Smith, T.C. Livestock-Associated MRSA and its current evolution. *Curr. Clin. Microbiol. Rep.* **2016**, *3*, 19–31. [CrossRef]
13. Cuny, C.; Abdelbary, M.; Layer, F.; Werner, G.; Witte, W. Prevalence of the immune evasion gene cluster in *Staphylococcus aureus* CC398. *Vet. Microbiol.* **2015**, *177*, 219–223. [CrossRef] [PubMed]
14. Jaja, I.F.; Green, E.; Muchenje, V. Aerobic mesophilic, coliform, *Escherichia coli,* and *Staphylococcus aureus* counts of raw meat from the formal and informal meat sectors in South Africa. *Int. J. Environ. Res. Public Health* **2018**, *15*, 819. [CrossRef] [PubMed]
15. Pekana, A.; Green, E. Antimicrobial resistance profiles of *Staphylococcus aureus* isolated from meat carcasses and bovine milk in abattoirs and dairy farms of the Eastern Cape, South Africa. *Int. J. Environ. Res. Public Health* **2018**, *15*, 2223. [CrossRef]
16. Tanih, N.F.; Sekwadi, E.; Ndip, R.N.; Bessong, P.O. Detection of pathogenic *Escherichia coli* and *Staphylococcus aureus* from cattle and pigs slaughtered in abattoirs in Vhembe District, South Africa. *J. World Sci.* **2015**, *2015*, 195972. [CrossRef]
17. Dweba, C.C.; Zishiri, O.T.; El Zowalaty, M.E. Isolation and molecular identification of virulence, antimicrobial and heavy metal resistance genes in livestock-associated methicillin-resistant *Staphylococcus aureus*. *Pathogens* **2019**, *8*, 79. [CrossRef]
18. Statistics South Africa. Formal Census 2011. Available online: https://www.statssa.gov.za/?page_id=3839 (accessed on 15 July 2020).
19. King Cetshwayo District Municipality (DC28). In *The Local Government Handbook: South Africa 2022*, 20th ed.; Main, O. (Ed.) Yes! Media: Mowbray, South Africa, 2022; p. 109. Available online: https://municipalities.co.za/map/124/king-cetshwayo-district-municipality (accessed on 1 November 2020).

20. Suresh, K.; Chandrashekara, S. Sample size estimation and power analysis for clinical research studies. *J. Human Rep. Sci.* **2012**, *5*, 7. [CrossRef]
21. Naing, L.; Winn, T.; Rusli, B.N. Practical issues in calculating the sample size for prevalence studies. *J. Arch. Sci.* **2006**, *1*, 9–14.
22. Lenth, R. Some practical guidelines for effective sample size determination. *J. Am. Stats.* **2001**, *55*, 187–193. [CrossRef]
23. Madoroba, E.; Magwedere, K.; Chaora, N.S.; Matle, I.; Muchadeyi, F.; Mathole, M.A.; Pierneef, R. Microbial communities of meat and meat roducts: An exploratory analysis of the product quality and safety at selected enterprises in South Africa. *Microorganisms* **2021**, *9*, 507. [CrossRef]
24. *ISO 6888-1*; Microbiology of Food and Animal Feeding Stuffs—Horizontal Method for Enumeration of Coagulase-Positive Staphylococci (Staphylococcus aureus and Other Species)—Part 1: Technique Using Baird-Parker Agar. ISO: Geneva, Switzerland, 1999; pp. 1–11.
25. Goja, A.; Ahmed, T.; Saeed, S.; Dirar, H. Isolation and identification of *Staphylococcus spp.* in fresh beef. *Pak. J. Nutr.* **2013**, *12*, 114. [CrossRef]
26. Govender, V.; Madoroba, E.; Magwedere, K.; Fosgate, G.; Kuonza, L. Prevalence and risk factors contributing to antibiotic-resistant *Staphylococcus aureus* isolates from poultry meat products in South Africa, 2015–2016. *J. S. Afr. Vet. Assoc.* **2019**, *90*, e1–e8. [CrossRef] [PubMed]
27. Bartholomew, J.W.; Mittwer, T. The gram stain. *J. Bacteriol.* **1952**, *16*, 1–29. [CrossRef] [PubMed]
28. Cameron, M.; Perry, J.; Middleton, J.R.; Chaffer, M.; Lewis, J.; Keefe, G.P. Evaluation of MALDI-TOF mass spectrometry and a custom reference spectra expanded database for the identification of bovine-associated coagulase-negative staphylococci. *J. Dairy Sci.* **2018**, *101*, 590–595. [CrossRef]
29. Savini, V.; Paparella, A.; Serio, A.; Marrollo, R.; Carretto, E.; Fazii, P. *Staphylococcus pseudintermedius* for CAMP-test. *Int. J. Clin. Exp. Pathol.* **2014**, *7*, 1733.
30. Hanson, A. *CAMP Test Protocols. Microbe Library Curriculum*; Website of The American Society for Microbiology: Washington, DC, USA, 2006; Available online: http://www.microbelibrary.org (accessed on 7 January 2022).
31. Hudzicki, J. Kirby-Bauer disk diffusion susceptibility test protocol. *Am. Soc. Microbiol.* **2016**, *16*, 1–23.
32. Bauer, A.W. Antibiotic susceptibility testing by a standardized single disc method. *Am. J. Clin. Pathol.* **1966**, *45*, 149–158. [CrossRef]
33. Spolaczyk, R.; Harnack, K. Measuring System for Optically Determining Concentration of Turbid Liquid Samples. U.S. Patents US6803594B2, 12 October 2004.
34. Clinical and Laboratory Standard Institute. Performance standard for antimicrobial susceptibility testing. In *Nineteenth Informational Supplement (M100-S19)*, 27th ed.; Clinical and Laboratory Standards Institute: Wayne, PA, USA, 2017.
35. Arfatahery, N.; Davoodabadi, A.; Abedimohtasab, T. Characterization of toxin genes and antimicrobial susceptibility of *Staphylococcus aureus* isolates in fishery products in Iran. *J. Sci. Rep.* **2016**, *6*, 1–7. [CrossRef]
36. Mehrotra, M.; Wang, G.; Johnson, W.M. Multiplex PCR for detection of genes for *Staphylococcus aureus* enterotoxins, exfoliative toxins, toxic shock syndrome toxin 1, and methicillin resistance. *J. Clin. Microbiol.* **2000**, *38*, 1032–1035. [CrossRef]
37. Petróczki, F.M.; Pásztor, Á.; Szűcs, K.D.; Pál, K.; Kardos, G.; Albert, E.; Horváth, B.; Ungvári, E.; Béri, B.; Peles, F. Occurrence and characteristics of *Staphylococcus aureus* in a Hungarian dairy Farm during a control program. *Pathogens* **2021**, *10*, 104. [CrossRef]
38. Lima, M.C.; de Barros, M.; Scatamburlo, T.M.; Polveiro, R.C.; de Castro, L.K.; Guimarães, S.H.S.; da Costa, S.L.; da Costa, M.M.; Moreira, M.A.S. Profiles of *Staphyloccocus aureus* isolated from goat persistent mastitis before and after treatment with enrofloxacin. *BMC Microbiol.* **2020**, *20*, 127. [CrossRef] [PubMed]
39. Tema, A.T. Microbiological Characterization of Unpasteurized Sheep Milk. Doctoral Dissertation, University Of Debrecen, Debrecen, Hungary, 2021.
40. Magiorakos, A.-P.; Srinivasan, A.; Carey, R.B.; Carmeli, Y.; Falagas, M.; Giske, C.; Harbarth, S.; Hindler, J.; Kahlmeter, G.; Olsson-Liljequist, B. Multidrug-resistant, extensively drug-resistant and pandrug-resistant bacteria: An international expert proposal for interim standard definitions for acquired resistance. *J. Clin. Microbiol. Infect.* **2012**, *18*, 268–281. [CrossRef] [PubMed]
41. Bounar-Kechih, S.; Taha Hamdi, M.; Aggad, H.; Meguenni, N.; Cantekin, Z. Carriage Methicillin-Resistant *Staphylococcus aureus* in poultry and cattle in Northern Algeria. *Vet. Med. Int.* **2018**, *2018*, 4636121. [CrossRef] [PubMed]
42. Attien, P.; Sina, H.; Moussaoui, W.; Dadieacute, T.; Chabi, S.K.; Djeacute ni, T.; Bankole, H.S.; Kotchoni, S.O.; Edoh, V.; Preacute vost, G.; et al. Prevalence and antibiotic resistance of *Staphylococcus* strains isolated from meat products sold in Abidjan streets (Ivory Coast). *Afr. J. Microbiol. Res.* **2013**, *7*, 3285–3293. [CrossRef]
43. El Tawab, A.A.A.; Maarouf, A.A.; El-Rais, E.M. Bacteriological and molecular studies on antibiotic resistant *Staphylococcus aureus* isolated from meat and its products in Qaliobaya, Egypt. *Vet. Med. J.* **2018**, *34*, 360–373.
44. Bissong, M.E.A.; Tahnteng, B.F.; Ateba, C.N.; Akoachere, J.T.K. Pathogenic potential and antimicrobial resistance profile of *Staphylococcus aureus* in milk and beef from the Northwest and Southwest Regions of Cameroon. *BioMed Res. Int.* **2020**, *2020*, 6015283. [CrossRef]
45. Adesiji, Y.O.; Alli, O.T.; Adekanle, M.A.; Jolayemi, J.B. Prevalence of *Arcobacter*, *Escherichia coli*, *Staphylococcus aureus* and *Salmonella* species in retail raw chicken, pork, beef and goat meat in Osogbo, Nigeria. *J. Biomed. Res.* **2011**, 3, 8–12. [CrossRef]
46. Aydin, A.; Sudagidan, M.; Muratoglu, K. Prevalence of staphylococcal enterotoxins, toxin genes and genetic-relatedness of foodborne *Staphylococcus aureus* strains isolated in the Marmara Region of Turkey. *Int. J. Food Microbiol.* **2011**, *148*, 99–106. [CrossRef]

47. Heo, H.J.; Ku, B.K.; Bae, D.H.; Park, C.K.; Lee, Y. Antimicrobial resistance of *Staphylococcus aureus* isolated from domestic and imported raw meat in Korea. *J. Vet. Res.* **2008**, *48*, 75–81.
48. Quddoumi, S.S.; Bdour, S.M.; Mahasneh, A.M. Isolation and characterization of methicillin-resistant *Staphylococcus aureus* from livestock and poultry meat. *Ann. Microbiol.* **2006**, *56*, 155–161. [CrossRef]
49. Thapaliya, D.; Forshey, B.M.; Kadariya, J.; Quick, M.K.; Farina, S.; O'Brien, A.; Nair, R.; Nworie, A.; Hanson, B.; Kates, A.; et al. Prevalence and molecular characterization of *Staphylococcus aureus* in commercially available meat over a one-year period in Iowa, USA. *Food Microbiol.* **2017**, *65*, 122–129. [CrossRef] [PubMed]
50. Pesavento, G.; Ducci, B.; Comodo, N.; Nostro, A.L. Antimicrobial resistance profile of *Staphylococcus aureus* isolated from raw meat: A research for methicillin resistant *Staphylococcus aureus* (MRSA). *Food Control* **2007**, *18*, 196–200. [CrossRef]
51. Abdalrahman, L.S.; Wells, H.; Fakhr, M.K. *Staphylococcus aureus* is more prevalent in retail beef livers than in pork and other beef cuts. *Pathogens* **2015**, *4*, 182–198. [CrossRef]
52. De Boer, E.; Zwartkruis-Nahuis, J.; Wit, B.; Huijsdens, X.; De Neeling, A.; Bosch, T.; Van Oosterom, R.; Vila, A.; Heuvelink, A.E. Prevalence of methicillin-resistant *Staphylococcus aureus* in meat. *Inter. J. Food Microbiol.* **2009**, *134*, 52–56. [CrossRef] [PubMed]
53. Lee, A.S.; de Lencastre, H.; Garau, J.; Kluytmans, J.; Malhotra-Kumar, S.; Peschel, A.; Harbarth, S. Methicillin-resistant *Staphylococcus aureus*. *Nat. Rev. Dis. Primers* **2018**, *4*, 1–23. [CrossRef]
54. Lutz, J.K.; Van Balen, J.; Mac Crawford, J.; Wilkins III, J.R.; Lee, J.; Nava-Hoet, R.C.; Hoet, A.E. Methicillin-resistant *Staphylococcus aureus* in public transportation vehicles (buses): Another piece to the epidemiologic puzzle. *Am. J. Infect. Control* **2014**, *42* 1285–1290. [CrossRef]
55. Alvseike, O.; Prieto, M.; Bjørnstad, P.H.; Mason, A. Intact gastro-intestinal tract removal from pig carcasses in a novel Meat Factory Cell approach. *J. Vet. Scan.* **2020**, *62*, 1–5. [CrossRef]
56. Burfoot, D.; Everis, L.; Mulvey, L.; Wood, A.; Campden, R.B. *Literature Review on Microbiological Hazards Associated with Biltong and Similar Dried Meat Products*; Report to Food Standard Agency: London, UK, 2010; pp. 1–87.
57. Tshipamba, M.E.; Lubanza, N.; Adetunji, M.C.; Mwanza, M. Molecular characterization and antibiotic resistance of foodborne pathogens in street-vended ready-to-eat meat sold in South Africa. *J. Food Prot.* **2018**, *81*, 1963–1972. [CrossRef] [PubMed]
58. Department of Health. *Guidelines for Environmental Health Officers on the Interpretation of Microbiological Analysis Data of Food*, Department of Health, Directorate Food Control: Pretoria, South Africa, 2000.
59. Huang, E.; Gurzau, A.E.; Hanson, B.M.; Kates, A.E.; Smith, T.C.; Pettigrew, M.M.; Spinu, M.; Rabinowitz, P.M. Detection of livestock-associated methicillin-resistant *Staphylococcus aureus* among swine workers in Romania. *J. Infect. Public Health* **2014**, *7*, 323–332. [CrossRef]
60. Loncaric, I.; Kuebber-Heiss, A.; Posautz, A.; Ruppitsch, W.; Lepuschitz, S.; Schauer, B.; Fessler, A.T.; Krametter-Froetscher, R.; Harrison, E.M.; Holmes, M.A. Characterization of *mecC* gene-carrying coagulase-negative *Staphylococcus* spp. isolated from various animals. *J. Vet. Microbiol.* **2019**, *230*, 138–144. [CrossRef]
61. Abolghait, S.K.; Fathi, A.G.; Youssef, F.M.; Algammal, A.M. Methicillin-resistant Staphylococcus aureus (MRSA) isolated from chicken meat and giblets often produces staphylococcal enterotoxin B (SEB) in non-refrigerated raw chicken livers. *Int. J. Food Microbiol.* **2020**, *328*, 108669. [CrossRef] [PubMed]
62. Normanno, G.; La Salandra, G.; Dambrosio, A.; Quaglia, N.; Corrente, M.; Parisi, A.; Santagada, G.; Firinu, A.; Crisetti, E.; Celano, G.V. Occurrence, characterization and antimicrobial resistance of enterotoxigenic *Staphylococcus aureus* isolated from meat and dairy products. *Int. J. Food Microbiol.* **2007**, *115*, 290–296. [CrossRef] [PubMed]
63. Nworie, A.; Onyema, A.S.; Okekpa, S.I.; Elom, M.O.; Umoh, N.O.; Usanga, V.U.; Ibiam, G.A.; Ukwah, B.N.; Nwadi, L.C.; Ezeruigbo, C.; et al. A novel Methicillin-Resistant *Staphylococcus aureus* t11469 and a poultry endemic strain t002 (ST5) are present in chicken in Ebonyi State, Nigeria. *BioMed Res. Int.* **2017**, *2017*, 2936461. [CrossRef] [PubMed]
64. Krupa, P.; Bystroń, J.; Bania, J.; Podkowik, M.; Empel, J.; Mroczkowska, A. Genotypes and oxacillin resistance of *Staphylococcus aureus* from chicken and chicken meat in Poland. *Poult. Sci.* **2014**, *93*, 3179–3186. [CrossRef]
65. Papadopoulos, P.; Papadopoulos, T.; Angelidis, A.S.; Boukouvala, E.; Zdragas, A.; Papa, A.; Hadjichristodoulou, C.; Sergelidis, D. Prevalence of *Staphylococcus aureus* and of methicillin-resistant *S. aureus* (MRSA) along the production chain of dairy products in north-western Greece. *Food Microbiol.* **2018**, *69*, 43–50. [CrossRef]
66. Bergdoll, M.S. Enterotoxins. In *Staphylococci and Staphylococcal Infections*; Easmon, C.S.F., Adlam, C., Eds.; Academic Press, Ltd.: London, UK, 1983; pp. 559–598.
67. Ikeda, T.; Tamate, N.; Yamaguchi, K.; Makino, S. Mass outbreak of food poisoning disease caused by small amounts of staphylococcal enterotoxins A and H. *Appl. Environ. Microbiol.* **2005**, *71*, 2793–2795. [CrossRef]
68. Evenson, M.L.; Hinds, M.W.; Bernstein, R.S.; Bergdoll, M.S. Estimation of human dose of staphylococcal enterotoxin A from a large outbreak of staphylococcal food poisoning involving chocolate milk. *Inter. J. Food Microbiol.* **1988**, *7*, 311–316. [CrossRef]
69. Fiebelkorn, K.R.; Crawford, S.A.; McElmeel, M.L.; Jorgensen, J.H. Practical disk diffusion method for detection of inducible clindamycin resistance in *Staphylococcus aureus* and coagulase-negative staphylococci. *J. Clin. Microbiol.* **2003**, *41*, 4740–4744. [CrossRef]
70. Hwang, S.Y.; Kim, S.H.; Jang, E.J.; Kwon, N.H.; Park, Y.K.; Koo, H.C.; Jung, W.K.; Kim, J.M.; Park, Y.H. Novel multiplex PCR for the detection of the *Staphylococcus aureus* superantigen and its application to raw meat isolates in Korea. *Int. J. Food Microbiol.* **2007**, *117*, 99–105. [CrossRef]

71. McLauchlin, J.; Narayanan, G.; Mithani, V.; O'neill, G. The detection of enterotoxins and toxic shock syndrome toxin genes in *Staphylococcus aureus* by polymerase chain reaction. *J. Food Prot.* **2000**, *63*, 479–488. [CrossRef]
72. Nashev, D.; Toshkova, K.; Bizeva, L.; Akineden, Ö.; Lämmler, C.; Zschöck, M. Distribution of enterotoxin genes among carriage- and infection-associated isolates of *Staphylococcus aureus*. *Lett. Appl. Microbiol.* **2007**, *45*, 681–685. [CrossRef] [PubMed]

microorganisms

MDPI

Communication

Food Is Reservoir of MDR *Salmonella*: Prevalence of ESBLs Profiles and Resistance Genes in Strains Isolated from Food

Delia Gambino †, Valeria Gargano *, Gaspare Butera †, Sonia Sciortino, Mariangela Pizzo, Giuseppa Oliveri, Cinzia Cardamone, Chiara Piraino, Giovanni Cassata, Domenico Vicari and Antonella Costa

Istituto Zooprofilattico Sperimentale della Sicilia "A. Mirri", 90129 Palermo, Italy; deliagamb@gmail.com (D.G.); gaspare.b@alice.it (G.B.); sonia.sciortino77@gmail.com (S.S.); mariangela.pizzo22@gmail.com (M.P.); giuseppa.oliveri@izssicilia.it (G.O.); cinzia.cardamone@izssicilia.it (C.C.); chiara.piraino@izssicilia.it (C.P.); giovanni.cassata@izssicilia.it (G.C.); domenico.vicari@izssicilia.it (D.V.); antonella.costa@izssicilia.it (A.C.)

* Correspondence: valeria.gargano@izssicilia.it

† These authors contributed equally to this work.

Abstract: *Salmonella* spp. are among the most frequent causes of foodborne diseases, and the increasing occurrence of MDR strains is an additional cause for concern. In the three-year period 2019–2021, we collected *Salmonella* spp. strains isolated from different food categories analysed in the context of Regulation (EC) No 2073/2005 in order to assess their antibiotic susceptibility profiles and ESBL production. To determine the susceptibility profiles and identify MDR strains, we used the Kirby–Bauer method to test 17 antibiotics. Double-disc and PCR testing then allowed us to assess the production of ESBLs and the presence of beta-lactamase resistance genes. Phenotypic tests showed that 36 out of 67 strains were MDR and 52.7% of these were ESBL producers. Finally, molecular investigations conducted on ESBL-producing strains revealed the presence of bla_{SHV}, $bla_{CTX\text{-}M}$ and bla_{TEM} genes. Our results confirmed the prevalence of *S. Infantis*, an MDR strain and ESBL producer, in chicken meat. This suggests that further research on the prevalence of antibiotic resistance genes (ARGs) in foodborne strains is needed, especially from a One Health perspective.

Keywords: *Salmonella*; food pathogens; *S. Infantis*; antibiotic resistance; resistance gene; ESBLs

Citation: Gambino, D.; Gargano, V.; Butera, G.; Sciortino, S.; Pizzo, M.; Oliveri, G.; Cardamone, C.; Piraino, C.; Cassata, G.; Vicari, D.; et al. Food Is Reservoir of MDR *Salmonella*: Prevalence of ESBLs Profiles and Resistance Genes in Strains Isolated from Food. *Microorganisms* **2022**, *10*, 780. https://doi.org/10.3390/microorganisms10040780

Academic Editor: Abiodun Adewale Adesiyun

Received: 22 March 2022
Accepted: 4 April 2022
Published: 6 April 2022

1. Introduction

Salmonellosis is a commonly reported gastrointestinal infection in humans, and an important cause of foodborne outbreaks. In the European Union (EU) in 2019, the number of confirmed salmonellosis cases was 87,923; in 2020, the number was 57,702, which was the lowest recorded number since 2007 because of the impacts of the withdrawal of the United Kingdom from the EU and the COVID-19 pandemic [1]. The main route of infection is ingestion of food or water contaminated with *Salmonella* spp., Gram-negative, facultative anaerobic bacilli belonging to the *Enterobacteriaceae* family [2,3]. *Salmonella* is ubiquitous in the human food chain and is one of the most important foodborne pathogens in the world. In particular, *S. Enteritidis*, *S. Typhimurium*, monophasic *S. Typhimurium*, *S. Infantis* and *S. Derby* are the five serotypes most commonly involved in human infections [1]. In the EU, microbiological food controls carried out in the context of Regulation (EC) No. 2073/2005 found the highest percentages of *Salmonella*-positive samples in egg products, poultry meat and poultry products, which are the most critical sources of *Salmonella* spp. transmission to humans [1,4].

Although salmonellosis is generally self-limited and usually does not require specific treatment, antibiotic therapy with quinolones, beta-lactams, aminoglycosides, tetracyclines or sulfamethoxazole–trimethoprim is necessary in severe cases [5]. However, the overuse of antibiotics has contributed to the selection of MDR *Salmonella* strains, i.e., resistant simultaneously to three or more classes of antibiotics, including those most commonly prescribed for the treatment of salmonellosis [6]. The spread of MDR *Salmonella* represents a significant health problem, as it causes longer hospitalisations, prolonged illnesses and higher mortality rates than susceptible strains [7,8]. The World Health Organization estimates that of the 100,000 cases of salmonellosis each year, a large number are caused by MDR *Salmonella* [9], with the majority acquired through the consumption of contaminated food of animal origin, particularly beef, pork and poultry products [10,11].

In *Enterobacteriaceae* such as *Salmonella*, the main mechanism of resistance to beta lactams is the acquisition of genes (*bla* gene) that encode for beta-lactamase hydrolytic enzymes, which inactivate the antibiotic [12]. Extended-spectrum beta-lactamases (ESBLs) which hydrolyse first-, second-, and third-generation penicillins and cephalosporins, are encoded by genes belonging to the TEM, SHV, and CTX-M families, including multiple variants of the bla_{TEM}, bla_{SHV} and $bla_{\mathrm{CTX\text{-}M}}$ genes [13]. These ESBL genes have been identified in bacteria isolated from animals and food products of animal origin [7,14], as well as from other types of foods, such as seafood [15], raw vegetables [16] and ready-to-eat (RTE) foods [17], suggesting the possible role of the food production chain as a reservoir for this group of bacteria. Indeed, factors such as selective pressure in animal and environmental microbiomes, the circulation of bacteria between animals and environment and ineffective food safety management can contribute to the presence and persistence of antibiotic-resistant bacteria (ARB) and antibiotic resistance genes (ARGs) in the food production context [18].

The aim of this study was to evaluate the MDR potential of *Salmonella* strains isolated in the period from January 2019 to December 2021 from food samples analysed in the context of the Regulation (EC) No 2073/2005 [4]. Furthermore, for every MDR *Salmonella* strain, ESBL production and ESBL gene presence were determined by double-disc diffusion and PCR tests, respectively.

2. Materials and Methods

2.1. *Salmonella* Isolation

From January 2019 to December 2021, 493 food samples, subjected to controls according to European Community legislation, were analysed [4]. Specifically, these samples were poultry meat (n = 145), pig meat (n = 106), beef (n = 54), bivalve molluscs (n = 109), eggs (n = 43) and sprouted seeds (n = 36).

Isolation according to ISO 6579-1:2017 was performed, and strains were then identified by biochemical enzymatic assays and serotyping, according to the Kauffmann–White–Le Minor scheme (Supplementary Materials Table S1) [19].

2.2. Antibiotic Susceptibility Profile Determination

Antibiotic susceptibility was assessed using the Kirby–Bauer method on Mueller Hinton agar medium (Oxoid, Milan, Italy), testing 17 antibiotics: kanamycin (30 μg), gentamicin (10 μg), streptomycin (10 μg), tobramycin (10 μg), ampicillin (10 μg), amoxicillin/clavulanic acid (30 μg), cefotaxime (30 μg), ceftriaxone (30 μg), ceftazidime (30 μg), imipenem (10 μg), nalidixic acid (30 μg), ciprofloxacin (5 μg), enrofloxacin (5 μg), levofloxacin (5 μg), sulfamethoxazole/trimethoprim (25 μg), tetracycline (30 μg) and chloramphenicol (30 μg).

Interpretation of inhibition zones and classification of isolates as susceptible (S), intermediate (I) or resistant (R), was done in accordance with CLSI guidelines [20].

2.3. ESBL Production Evaluation by Double-Disc Test

The double-disc test (DDT) was conducted on 36 MDR *Salmonella* strains to phenotypically assess ESBL production. Discs containing cephalosporins (cefotaxime 30 μg, ceftazidime 30 μg, cefepime 30 μg) were placed next to a disc with clavulanic acid (30 μg amoxicillin–clavulanic acid), as recommended by EUCAST [21]. When zones of inhibition around any of the cephalosporin discs were increased or there was a 'keyhole' in the direction of amoxicillin–clavulanic acid disc, the test was considered positive.

2.4. Detection of Beta-Lactamase Genes

The beta-lactamase gene detection was conducted on the 19 strains that were found by the double-disc test to be ESBL-producing. Bacterial DNA was extracted using 100 µL of PrepMan™ ultra Sample Preparation Reagent (Thermo Fisher Scientific, Waltham, MA, USA), according to the procedure recommended by the manufacturer. Real Time PCR reactions were performed using 10 ng of DNA template and 0.5 µM of the forward and reverse primers listed in Table 1, for a total volume of 25 µL of 1X of Advanced Universal SYBR Green Supermix (Bio-Rad Laboratories, Hercules, CA, USA), in order to amplify bla_{TEM}, $bla_{CTX\text{-}M}$, bla_{SHV} and bla_{OXA} genes.

Table 1. Primers used in this study.

Target	Primers	Sequence (5′-3′)	Amplicon Size (bp)	Reference
bla_{TEM}	bla_{TEM}_F bla_{TEM}_R	ATTCTTGAAGACGAAAGGGC ACGCTCAGTGGAACGAAAAC	661	[22]
$bla_{CTX\text{-}M}$	$bla_{CTX\text{-}M}$_F $bla_{CTX\text{-}M}$_R	CGCTTTGCGATGTGCAG ACCGCGATATCGTTGGT	585	
bla_{SHV}	bla_{SHV}_F bla_{SHV}_R	CACTCAAGGATGTATTGTG TTAGCGTTGCCAGTGCTCG	807	
bla_{OXA}	bla_{OXA}_F bla_{OXA}_R	ACACAATACATATCAACTTCGC AGTGTGTTTAGAATGGTGATC	590	

The amplification program included an initial denaturation at 94 °C for 10 min, followed by 32 cycles of 94 °C for 30 s, 60 °C for 30 s, 72 °C for 15 s, and a final extension at 72 °C for 10 min. Subsequently, 10 µL of the PCR product were used for electrophoresis on 2% E-Gel™ Go! Agarose Gels (Thermo Fisher Scientific, Waltham, MA, USA) to determine the size of the product. In each Real Time PCR reaction, a positive and a negative control were used. The positive one was represented by DNA belonging to a strain of *Salmonella* in which the presence of the *bla* gene was previous confirmed by sequencing; the negative control was represented by a Not Template Control (NTC), in which the reaction volume with DNase free water was obtained.

3. Results

3.1. Isolation Results

Microbiological analysis of the 493 food samples resulted in the isolation of 67 strains of *Salmonella* spp. (15 out of 172 were isolated in 2019, 17 out of 132 in 2020 and 35 out of 189 in 2021). Supplementary Materials Table S1 shows the samples that tested positive for the presence of *Salmonella* spp. and the serotypes identified. Notably, poultry meat was the main source of *Salmonella*, showing a prevalence of 40%, 52.9% and 71.4% in 2019, 2020 and 2021, respectively (Figure 1).

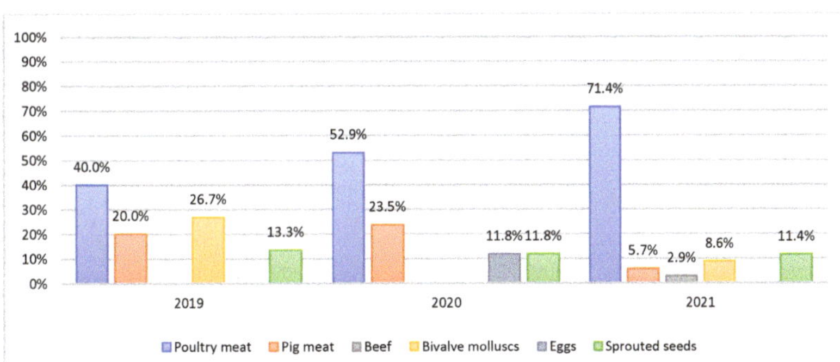

Figure 1. Prevalence per year of *Salmonella* based on food.

S. Infantis was the predominant serotype (48%), present in 32 poultry meat samples. Instead, *S. Typhimurium* (9%), *S. Derby* (6%) and *S. Enteritidis* (3%) serotypes were found to have a low prevalence (Figure 2).

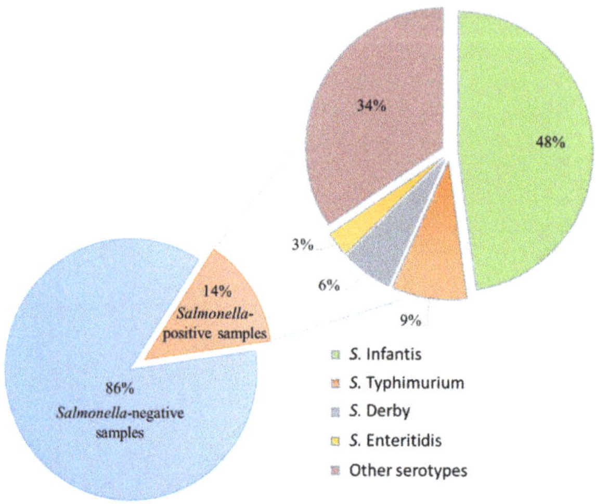

Figure 2. *Salmonella* spp. research results and serotypes identified in the 2019–2021 three-year period.

3.2. Antibiotic Susceptibility and ESBL Production Test Results

Antibiotic susceptibility testing conducted on the 67 *Salmonella* strains showed the absence of resistance in 24 of these strains, whereas 43 strains (64%) were resistant to one or more of the tested antibiotics. Supplementary Materials Table S1 provides an overview of these strains and their resistances.

Notably, 31.3% of these strains were resistant to kanamycin, 43.2% to sulphonamides, 47.7% to nalidixic acid, 49.2% to ampicillin and 50.7% to tetracycline. Few strains showed resistance to levofloxacin (5%) or chloramphenicol (6%), whereas no resistance against imipenem, ciprofloxacin or enrofloxacin was detected.

An MDR profile was found in 36 strains that showed resistance to three (n = 4), four (n = 22) and five (n = 10) antibiotic classes (Supplementary Materials Table S1). Specifically, the most frequent MDR profiles were: aminoglycosides, beta-lactams, quinolones, sulphonamides and tetracyclines; resistance to these was found in eight *S. Infantis*, one *S. Salamae* and one *S. Kentucky*. Resistance to beta-lactams, quinolones, sulphonamides and tetracyclines was found in nine *S. Infantis* and one *S. Cerro*.

Finally, the double-disc test allowed detection of ESBL production in 19 strains. Indeed, for these strains, an increase in the zones of inhibition in the direction of amoxicillin or clavulanic acid was recorded around the tested cephalosporins (Table 2).

Table 2. Resistance and ESBL production test results of the 36 MDR *Salmonella* strains.

ID	Food	*Salmonella* Serotype	Isolation Year	Resistance	ESBL Production
AL-3	Poultry meat	*S. Infantis*	2019	AMP, CTX, NAL, SXT, TET	–
AL-11	Poultry meat	*S. Newport*	2019	KAN, AMP, SXT, TET	+
AL-14	Poultry meat	*S. Infantis*	2019	KAN, AMP, CTX, NAL, TET	–
AL-15	Poultry meat	*S. Infantis*	2019	KAN, AMP, CTX, NAL, SXT, TET	–
AL-30	Poultry meat	*S. Infantis*	2020	KAN, NAL, SXT, TET	+
AL-20	Poultry meat	*S. Infantis*	2020	NAL, SXT, TET	+
AL-21	Poultry meat	*S. Infantis*	2020	KAN, NAL, SXT, TET	+
AL-25	Poultry meat	*S. Infantis*	2020	STR, AMP, NAL, SXT, TET	+
AL-26	Poultry meat	*S. Infantis*	2020	STR, NAL, SXT, TET	+
AL-27	Poultry meat	*S. Infantis*	2020	KAN, STR, NAL, SXT, TET	+
AL-32	Poultry meat	*S. Infantis*	2020	KAN, STR, NAL, SXT, TET	+
AL-34	Poultry meat	*S. Infantis*	2021	KAN, STR, AMP, CTX, NAL, LEVO, CHL	–

Table 2. *Cont.*

ID	Food	*Salmonella* Serotype	Isolation Year	Resistance	ESBL Production
AL-35	Poultry meat	*S.* Infantis	2021	KAN, STR, AMP, CTX, NAL, LEVO, CHL	−
AL-37	Poultry meat	*S.* Agona	2021	STR, AMP, SXT	+
AL-38	Pig meat	*S.* Salamae	2021	KAN, GEN, TOB, AMP, AMC, NAL, SXT, CHL	+
AL-39	Poultry meat	*S.* Infantis	2021	KAN, SXT, TET	+
AL-43	Poultry meat	*S.* Infantis	2021	KAN, AMP, STR, NAL, SXT, TET	+
AL-44	Poultry meat	*S.* Infantis	2021	KAN, TOB, AMP, CTX, CRO, NAL, SXT	−
AL-45	Poultry meat	*S.* Infantis	2021	AMP, CTX, CRO, NAL, SXT, TET	−
AL-46	Poultry meat	*S.* Infantis	2021	STR, AMP, NAL, TET	
AL-47	Beef	*S.* Cerro	2021	AMP, AMC, CTX, CRO, NAL, SXT, TET	−
AL-48	Poultry meat	*S.* Infantis	2021	KAN, GEN, TOB, AMP, AMC, CTX, CRO, NAL, SXT, TET	−
AL-49	Poultry meat	*S.* Infantis	2021	KAN, AMP, AMC, CTX, CRO, NAL, SXT, TET	−
AL-50	Poultry meat	*S.* Infantis	2021	KAN, AMP, AMC, CTX, CRO, SXT, TET	−
AL-51	Pig meat	*S.* Typhimurium	2021	STR, AMP, TET	+
AL-52	Poultry meat	*S.* Infantis	2021	AMP, CTX, CAZ, CRO, NAL, SXT, TET	+
AL-53	Poultry meat	*S.* Kentucky	2021	STR, AMP, CAZ, CTX, CRO, NAL, SXT, TET	+
AL-56	Poultry meat	*S.* Infantis	2021	AMP, AMC, CRO, NAL, SXT, TET	−
AL-57	Poultry meat	*S.* Infantis	2021	TOB, AMP, AMC, CTX, CRO, NAL, SXT, TET	+
AL-58	Poultry meat	*S.* Infantis	2021	AMP, AMC, CTX, NAL, SXT, TET	+
AL-59	Poultry meat	*S.* Infantis	2021	GEN, AMP, CTX, NAL, SXT, TET	−
AL-60	Poultry meat	*S.* Infantis	2021	KAN, TOB, AMP, CTX, NAL, SXT, TET	+
AL-63	Poultry meat	*S.* Infantis	2021	STR, AMP, CTX, CAZ, NAL, SXT, TET, CHL	+
AL-65	Poultry meat	*S.* Infantis	2021	AMP, CTX, CAZ, CRO, NAL, SXT, TET	−
AL-66	Poultry meat	*S.* Infantis	2021	KAN, STR, AMP, NAL, TET	−
AL-67	Poultry meat	*S.* Infantis	2021	KAN, AMP, NAL, TET	−

AMP, Ampicillin; CTX, Cefotaxime; NAL, Nalidixic Acid; SXT, Sulphamethoxazole/Trimethoprim; TET, Tetracycline; KAN, Kanamycin; GEN, Gentamicin; STR, Streptomycin; TOB, Tobramycin; AMC, Amoxicillin/Clavulanic acid; CAZ, Ceftazidime; CRO, Ceftriaxone; LEVO, Levofloxacin; CHL, Chloramphenicol.

3.3. Detection of Beta-Lactamase Genes

Genes responsible for beta-lactamase activity in 19 ESBL-producing *Salmonella* strains were screened by PCR. The presence of beta-lactamase genes was detected in all tested strains, confirming the phenotypic results of ESBL production tests (Table 3).

Table 3. Beta-lactamase resistance gene detection results.

ID Strains	Food	*Salmonella* Serotype	ESBL Production	*bla* Gene Detected
AL-11	Poultry meat	*S.* Newport	+	bla_{TEM}, bla_{SHV}
AL-30	Poultry meat	*S.* Infantis	+	$bla_{CTX\text{-}M}$
AL-20	Poultry meat	*S.* Infantis	+	bla_{SHV}
AL-21	Poultry meat	*S.* Infantis	+	$bla_{CTX\text{-}M}$
AL-25	Poultry meat	*S.* Infantis	+	bla_{SHV}
AL-26	Poultry meat	*S.* Infantis	+	bla_{SHV}
AL-27	Poultry meat	*S.* Infantis	+	$bla_{CTX\text{-}M}$
AL-32	Poultry meat	*S.* Infantis	+	bla_{SHV}
AL-37	Poultry meat	*S.* Agona	+	bla_{SHV}
AL-38	Pig meat	*S.* Salamae	+	$bla_{CTX\text{-}M}$, bla_{SHV}
AL-39	Poultry meat	*S.* Infantis	+	bla_{SHV}
AL-43	Poultry meat	*S.* Infantis	+	$bla_{CTX\text{-}M}$
AL-51	Pig meat	*S.* Typhimurium	+	bla_{SHV}
AL-52	Poultry meat	*S.* Infantis	+	bla_{SHV}
AL-53	Poultry meat	*S.* Kentucky	+	$bla_{CTX\text{-}M}$
AL-57	Poultry meat	*S.* Infantis	+	$bla_{CTX\text{-}M}$
AL-58	Poultry meat	*S.* Infantis	+	$bla_{CTX\text{-}M}$, bla_{SHV}
AL-60	Poultry meat	*S.* Infantis	+	bla_{SHV}
AL-63	Poultry meat	*S.* Infantis	+	$bla_{CTX\text{-}M}$, bla_{SHV}

The most frequently identified genes were bla_{SHV} and $bla_{CTX\text{-}M}$, which were present in 68.4% and 47.3% of strains, respectively. Furthermore, the bla_{TEM} gene was harboured by only one strain, while bla_{OXA} was not detected. Specifically, nine strains harboured only the bla_{SHV} gene, six strains harboured only the $bla_{CTX\text{-}M}$ gene, three strains harboured the $bla_{CTX\text{-}M}$ and bla_{SHV} genes together, and one strain harboured the bla_{TEM} and bla_{SHV} genes together.

4. Discussion

Salmonella spp. are among the most frequent causes of foodborne diseases, and the increasing occurrence of MDR strains is an additional cause for concern. Thus, in the three-year period 2019–2021, we collected *Salmonella* spp. strains isolated from different food categories analysed in the context of Regulation (EC) No 2073/2005 [4], in order to assess their antibiotic susceptibility profiles and ESBL production.

Our data show that among the different food categories analysed, poultry meat was a relevant source of *Salmonella*. Moreover, regarding poultry meat, it is possible to note that the prevalence of *Salmonella* significantly increased over the three-year period, rising from 40% in 2019 to 71.4% in 2021; the prevalent serovar was *S. Infantis* (48%).

We performed a screening test using the Kirby–Bauer method to estimate the antibiotic susceptibility profiles of these strains, and we found a very high rate of strains showing at least one phenotypic resistance (64%). Among these, the highest rates of resistance were found against sulphonamides (43.2%), a class of antibiotics used in severe *Salmonella* infections, but also against nalidixic acid (47.7%) and kanamycin (31.3%). In addition, a high percentage of strains showed resistance to tetracyclines (50.7%), despite the fact that in 2006, the European Union, in an attempt to counteract this trend, imposed a ban on the non-therapeutic use of antibiotics of human importance, such as tetracyclines, in farm animal feed. However, resistance to these drugs in *Salmonella* from food samples continues to be of concern [8,23]. This observation may be related to the human manipulation of these kinds of foods [24].

Of the strains tested, 53.7% showed an MDR profile with resistance to four or five classes in the majority of strains. These data are alarming, not only because of the real risk for consumers of becoming infected with an MDR strain, but also because many of these strains showed resistance to antibiotic classes important in human medicine such as beta-lactamases. Thus, in order to obtain a complete overview of the resistance profiles of all the MDR strains isolated, we conducted a double-disc test (DDT) for ESBL phenotype detection. This test is one of the four different methods for confirming the ESBL phenotype recommended by EUCAST [21]. Despite the EFSA 2018/2019 report's observation of resistance to third-generation cephalosporins at the overall low levels of 1.8% and 1.2% for cefotaxime and ceftazidime, respectively, for *Salmonella* spp., our experiment indicated that 52% of all MDR strains had an ESBL phenotype [8]. Finally, because these phenotypes could be conferred by several ARGs [25], the detection of beta-lactamase genes was performed in order to confirm phenotypic pattern. The PCRs we conducted allowed us to identify at least one gene encoding for β-lactamase enzymes in each strain that had an ESBL profile (Table 3). The $bla_{\text{CTX-M}}$ gene was present in 9 out of 19 ESBL strains, and in three of these, it was in association with the bla_{SHV} gene, which was found to be the most prevalent gene among our isolates, because of its detection in 12 out of 19 ESBL strains. The $bla_{\text{CTX-M}}$ genes encode for extended-spectrum β-lactamases (ESBLs) frequently identified in Gram-negative pathogens. These types of enzymes are active against cephalosporins and monobactams (but not cephamycins or carbapenems), and are currently of great epidemiological and clinical interest [26]. The bla_{SHV} gene has been identified mainly in *Enterobacteriaceae* causing nocosomal infections, but also in isolates from different contexts (human, animal and environment) [27,28]. Probably originating from a chromosomal penicillinase of *Klebsiella pneumoniae*, SHV β-lactamases currently comprise a large number of allelic variants, including extended-spectrum β-lactamases (ESBLs), non-ESBLs and several unclassified variants [29]. Our isolates showed an ESBL phenotype, so we have probably identified bla_{SHV} genes encoded for extended-spectrum β-lactamases.

These data are certainly alarming, since all of our strains came from food samples, particularly poultry, intended for human consumption. Indeed, although cooking these products may reduce the risk of foodborne disease, ARGs can resist high temperatures and, once ingested, can be transferred to the gut microbiota and confer resistance to other bacteria [30]. Therefore, our data are in line with the latest EFSA recommendations, which confirm how important it is in the monitoring and surveillance of antibiotic resistance (AMR) to assess the presence of ARGs in foodborne strains, especially in a One Health approach that recognises the circularity of human, animal and environmental health.

Supplementary Materials: The following supporting information can be downloaded at: https://www.mdpi.com/article/10.3390/microorganisms10040780/s1. Table S1: Analysed strains and their phenotypic resistances.

Author Contributions: Conceptualisation, V.G., D.G. and A.C.; methodology, V.G., C.P., M.P., C.C. and D.G.; formal analysis, D.G., V.G. and G.B.; investigation, D.G., V.G., S.S., G.O. and G.B.; resources, A.C.; data curation, V.G. and D.G.; writing—original draft preparation, V.G., D.G. and G.B.; writing—review and editing, V.G., D.G., S.S., C.C. and A.C.; visualisation, D.V.; supervision, G.C., D.V. and A.C.; project administration, A.C.; funding acquisition, A.C. All authors have read and agreed to the published version of the manuscript.

Funding: This research was funded by IZS SI 07/20 RC: "Study on the correlation between pesticides and antimicrobial resistance in bacteria isolated from plants".

Institutional Review Board Statement: Not applicable.

Informed Consent Statement: Not applicable.

Data Availability Statement: All data discussed are contained in the manuscript.

Conflicts of Interest: The authors declare no conflict of interest.

References

1. European Food Safety Authority; European Centre for Disease Prevention and Control. The European Union One Health 2020 Zoonoses Report. *EFSA J.* **2021**, *19*, e06971. [CrossRef]
2. Liu, H.; Whitehouse, C.A.; Li, B. Presence and Persistence of *Salmonella* in Water: The Impact on Microbial Quality of Water and Food Safety. *Front. Public Health* **2018**, *6*, 159. [CrossRef] [PubMed]
3. Ehuwa, O.; Jaiswal, A.K.; Jaiswal, S. *Salmonella*, Food Safety and Food Handling Practices. *Foods* **2021**, *10*, 907. [CrossRef] [PubMed]
4. Commission of the European Communities Regulation (EC) No 2073/2005 of 15 November 2005 on Microbiological Criteria for Foodstuffs. 2005. Available online: http://data.europa.eu/eli/reg/2005/2073/oj (accessed on 14 December 2021).
5. Eng, S.-K.; Pusparajah, P.; Ab Mutalib, N.-S.; Ser, H.-L.; Chan, K.-G.; Lee, L.-H. *Salmonella*: A Review on Pathogenesis, Epidemiology and Antibiotic Resistance. *Front. Life Sci.* **2015**, *8*, 284–293. [CrossRef]
6. Elnekave, E.; Hong, S.L.; Lim, S.; Boxrud, D.; Rovira, A.; Mather, A.E.; Perez, A.; Alvarez, J. Transmission of Multidrug-Resistant *Salmonella* Enterica Subspecies Enterica 4,[5],12:I:- Sequence Type 34 between Europe and the United States. *Emerg. Infect. Dis.* **2020**, *26*, 3034–3038. [CrossRef] [PubMed]
7. Proietti, P.C.; Stefanetti, V.; Musa, L.; Zicavo, A.; Dionisi, A.M.; Bellucci, S.; Mensa, A.L.; Menchetti, L.; Branciari, R.; Ortenzi, R.; et al. Genetic Profiles and Antimicrobial Resistance Patterns of *Salmonella* Infantis Strains Isolated in Italy in the Food Chain of Broiler Meat Production. *Antibiotics* **2020**, *9*, 814. [CrossRef]
8. European Food Safety Authority; European Centre for Disease Prevention and Control. The European Union Summary Report on Antimicrobial Resistance in Zoonotic and Indicator Bacteria from Humans, Animals and Food in 2018/2019. *EFSA J.* **2021**, *19*, e06490. [CrossRef]
9. *Antimicrobial Resistance: Global Report on Surveillance*; World Health Organization: Geneva, Switzerland, 2014; ISBN 978-92-4-156474-8.
10. Castro-Vargas, R.E.; Herrera-Sánchez, M.P.; Rodríguez-Hernández, R.; Rondón-Barragán, I.S. Antibiotic Resistance in *Salmonella* Spp. Isolated from Poultry: A Global Overview. *Vet. World* **2020**, *13*, 2070–2084. [CrossRef]
11. Campos, J.; Mourão, J.; Peixe, L.; Antunes, P. Non-Typhoidal *Salmonella* in the Pig Production Chain: A Comprehensive Analysis of Its Impact on Human Health. *Pathogens* **2019**, *8*, 19. [CrossRef]
12. Iredell, J.; Brown, J.; Tagg, K. Antibiotic Resistance in Enterobacteriaceae: Mechanisms and Clinical Implications. *BMJ* **2016**, *352*, h6420. [CrossRef]
13. Philippon, A.; Slama, P.; Dény, P.; Labia, R. A Structure-Based Classification of Class A β-Lactamases, a Broadly Diverse Family of Enzymes. *Clin. Microbiol. Rev.* **2016**, *29*, 29–57. [CrossRef] [PubMed]
14. Ye, Q.; Wu, Q.; Zhang, S.; Zhang, J.; Yang, G.; Wang, J.; Xue, L.; Chen, M. Characterization of Extended-Spectrum β-Lactamase-Producing Enterobacteriaceae From Retail Food in China. *Front. Microbiol.* **2018**, *9*, 1709. [CrossRef] [PubMed]
15. Nguyen, D.T.A.; Kanki, M.; Nguyen, P.D.; Le, H.T.; Ngo, P.T.; Tran, D.N.M.; Le, N.H.; Dang, C.V.; Kawai, T.; Kawahara, R.; et al. Prevalence, Antibiotic Resistance, and Extended-Spectrum and AmpC β-Lactamase Productivity of *Salmonella* Isolates from Raw Meat and Seafood Samples in Ho Chi Minh City, Vietnam. *Int. J. Food Microbiol.* **2016**, *236*, 115–122. [CrossRef]

16. Colosi, I.A.; Baciu, A.M.; Opriș, R.V.; Peca, L.; Gudat, T.; Simon, L.M.; Colosi, H.A.; Costache, C. Prevalence of ESBL, AmpC and Carbapenemase-Producing *Enterobacterales* Isolated from Raw Vegetables Retailed in Romania. *Foods* **2020**, *9*, 1726. [CrossRef]
17. Kim, H.-S.; Chon, J.-W.; Kim, Y.-J.; Kim, D.-H.; Kim, M.; Seo, K.-H. Prevalence and Characterization of Extended-Spectrum-β-Lactamase-Producing Escherichia Coli and Klebsiella Pneumoniae in Ready-to-Eat Vegetables. *Int. J. Food Microbiol.* **2015**, *207*, 83–86. [CrossRef] [PubMed]
18. EFSA Panel on Biological Hazards (BIOHAZ); Koutsoumanis, K.; Allende, A.; Álvarez-Ordóñez, A.; Bolton, D.; Bover-Cid, S.; Chemaly, M.; Davies, R.; De Cesare, A.; Herman, L.; et al. Role Played by the Environment in the Emergence and Spread of Antimicrobial Resistance (AMR) through the Food Chain. *EFSA J.* **2021**, *19*, 6651. [CrossRef]
19. *ISO 6579-1:2017*; Microbiology of the Food Chain—Horizontal Method for the Detection, Enumeration and Serotyping of Salmonella—Part 1: Detection of Salmonella spp. International Organization for Standardization; AFNOR: Saint-Denis, France, 2017.
20. Clinical and Laboratory Standards Institute (CLSI). *Performance Standards for Antimicrobial Susceptibility: Supplement M100*, 31st ed.; Clinical and Laboratory Standards Institute: Wayne, PA, USA, 2021; ISBN 978-1-68440-105-5.
21. European Committee on Antimicrobial Susceptibility Testing (EUCAST). Guideline for the Detection of Resistance Mechanisms and Specific Resistances of Clinical and/or Epidemiological Importance, Version 2.0; 2017. Available online: https://www.eucast.org/resistance_mechanisms/ (accessed on 14 December 2021).
22. Zhao, X.; Hu, M.; Zhang, Q.; Zhao, C.; Zhang, Y.; Li, L.; Qi, J.; Luo, Y.; Zhou, D.; Liu, Y. Characterization of Integrons and Antimicrobial Resistance in *Salmonella* from Broilers in Shandong, China. *Poult. Sci.* **2020**, *99*, 7046–7054. [CrossRef] [PubMed]
23. Gargano, V.; Sciortino, S.; Gambino, D.; Costa, A.; Agozzino, V.; Reale, S.; Alduina, R.; Vicari, D. Antibiotic Susceptibility Profile and Tetracycline Resistance Genes Detection in *Salmonella* Spp. Strains Isolated from Animals and Food. *Antibiotics* **2021**, *10*, 809. [CrossRef] [PubMed]
24. Gargano, V.; Gambino, D.; Migliore, S.; Vitale, M.; Sciortino, S.; Costa, A.; Vicari, D. Can Human Handling Increase the Presence of Multidrug Resistance (MDR) in Salmonella Spp. Isolated from Food Sources? *Microorganisms* **2021**, *9*, 2018. [CrossRef] [PubMed]
25. Dhanani, A.S.; Block, G.; Dewar, K.; Forgetta, V.; Topp, E.; Beiko, R.G.; Diarra, M.S. Genomic Comparison of Non-Typhoidal Salmonella Enterica Serovars Typhimurium, Enteritidis, Heidelberg, Hadar and Kentucky Isolates from Broiler Chickens. *PLoS ONE* **2015**, *10*, e0128773. [CrossRef]
26. Cantón, R.; Coque, T.M. The CTX-M β-Lactamase Pandemic. *Curr. Opin. Microbiol.* **2006**, *9*, 466–475. [CrossRef] [PubMed]
27. Alcalá, L.; Alonso, C.A.; Simón, C.; González-Esteban, C.; Orós, J.; Rezusta, A.; Ortega, C.; Torres, C. Wild Birds, Frequent Carriers of Extended-Spectrum β-Lactamase (ESBL) Producing *Escherichia* Coli of CTX-M and SHV-12 Types. *Microb. Ecol.* **2016**, *72*, 861–869. [CrossRef] [PubMed]
28. Liakopoulos, A.; Mevius, D.; Ceccarelli, D. A Review of SHV Extended-Spectrum β-Lactamases: Neglected Yet Ubiquitous. *Front. Microbiol.* **2016**, *7*, 1374. [CrossRef] [PubMed]
29. Bush, K. Past and Present Perspectives on β-Lactamases. *Antimicrob. Agents Chemother.* **2018**, *62*, e01076-18. [CrossRef] [PubMed]
30. Groussin, M.; Poyet, M.; Sistiaga, A.; Kearney, S.M.; Moniz, K.; Noel, M.; Hooker, J.; Gibbons, S.M.; Segurel, L.; Froment, A.; et al. Elevated Rates of Horizontal Gene Transfer in the Industrialized Human Microbiome. *Cell* **2021**, *184*, 2053–2067.e18. [CrossRef] [PubMed]

microorganisms

MDPI

Communication

Identification and Subtyping of *Salmonella* Isolates Using Matrix-Assisted Laser Desorption–Ionization Time-of-Flight Mass Spectrometry (MALDI-TOF)

Anil K. Persad [1,2], Hanan A. Fahmy [1,3], Nicholas Anderson [1], Jing Cui [4], Zeynal Topalcengiz [5,6], Saharuetai Jeamsripong [7,8], Patrick M. Spanninger [9], Robert L. Buchanan [10], Kalmia E. Kniel [9], Michele T. Jay-Russell [7], Michelle D. Danyluk [6], Gireesh Rajashekara [1] and Jeffrey T. LeJeune [1,*,†]

1 Center for Food Animal Health, Department of Animal Sciences, Ohio Agricultural Research and Development Center, The Ohio State University, Wooster, OH 44691, USA; anilkpersad@gmail.com (A.K.P.); dr.hanana.fahmy@hotmail.com (H.A.F.); anderson.3052@osu.edu (N.A.); rajashekara.2@osu.edu (G.R.)
2 School of Veterinary Medicine, Faculty of Medical Sciences, The University of the West Indies, Eric Williams Medical Sciences Complex, Mt. Hope, Trinidad and Tobago, West Indies
3 Animal Health Research Institute (AHRI), Giza 12618, Egypt
4 Animal Disease Diagnostic Laboratory, Ohio Department of Agriculture, 8995 East Main St., Building 6, Reynoldsburg, OH 43068, USA; jing.cui@agri.ohio.gov
5 Department of Food Engineering, Faculty of Engineering and Architecture, Muş Alparslan University, Muş 49250, Turkey; zeynaltopalcengiz@gmail.com
6 Department of Food Science and Human Nutrition, Citrus Research and Education Center, University of Florida, Lake Alfred, FL 33850, USA; mddanyluk@ufl.edu
7 Western Institute for Food Safety and Security, UC Davis, Davis, CA 95618, USA; sjeamsripong@ucdavis.edu (S.J.); mjay@ucdavis.edu (M.T.J.-R.)
8 Research Unit in Microbial Food Safety and Antimicrobial Resistance, Department of Veterinary Public Health, Faculty of Veterinary Science, Chulalongkorn University, Pathumwan, Bangkok 10330, Thailand
9 Department of Animal and Food Sciences, University of Delaware, Newark, DE 19716, USA; pspannin@udel.edu (P.M.S.); kniel@udel.edu (K.E.K.)
10 Department of Nutrition and Food Science and Center for Food Safety and Security Systems, University of Maryland, College Park, MA 20742, USA; rbuchana@umd.edu
* Correspondence: jeffrey.lejeune@fao.org
† Current address: Food and Agriculture Organization of the United Nations (FAO), 00153 Rome, Italy.

Citation: Persad, A.K.; Fahmy, H.A.; Anderson, N.; Cui, J.; Topalcengiz, Z.; Jeamsripong, S.; Spanninger, P.M.; Buchanan, R.L.; Kniel, K.E.; Jay-Russell, M.T.; et al. Identification and Subtyping of *Salmonella* Isolates Using Matrix-Assisted Laser Desorption–Ionization Time-of-Flight Mass Spectrometry (MALDI-TOF). *Microorganisms* **2022**, *10*, 688. https://doi.org/10.3390/microorganisms10040688

Academic Editor: Abiodun Adewale Adesiyun

Received: 21 January 2022
Accepted: 16 March 2022
Published: 23 March 2022

Abstract: Subtyping of bacterial isolates of the same genus and species is an important tool in epidemiological investigations. A number of phenotypic and genotypic subtyping methods are available; however, most of these methods are labor-intensive and time-consuming and require considerable operator skill and a wealth of reagents. Matrix-Assisted Laser Desorption–Ionization Time-of-Flight Mass Spectrometry (MALDI-TOF), an alternative to conventional subtyping methods, offers a rapid, reproducible method for bacterial identification with a high sensitivity and specificity and at minimal cost. The purpose of this study was to determine the feasibility of using MALDI-TOF to differentiate between six *Salmonella* serovars recovered from experimental microcosms inoculated with known strains of *Salmonella*. Following the establishment of a MALDI-TOF reference library for this project, the identity of 843 *Salmonella* isolates recovered from these microcosms was assessed using both MALDI-TOF and conventional methods (serotyping/PCR). All 843 isolates were identified as being *Salmonella* species. Overall, 803/843 (95%) of these isolates were identified similarly using the two different methods. Positive percent agreement at the serovar level ranged from 79 to 100%, and negative percent agreement for all serovars was greater than 98%. Cohen's kappa ranged from 0.85 to 0.98 for the different serovars. This study demonstrates that MALDI-TOF is a viable alternative for the rapid identification and differentiation of *Salmonella* serovars.

Keywords: mass spectrometry; matrix-assisted laser desorption–ionization; *Salmonella*; subtyping; serology; MALDI-TOF

1. Introduction

Salmonella enterica serovars have been associated with foodborne disease globally for over 100 years [1]. While the global incidence is unknown, within the United States the disease burden is estimated to be over one million cases annually, with 16,000 hospitalizations and almost 500 deaths [2]. Most human cases are self-limiting; however, young children under five years, the elderly, and the immunocompromised are at the highest risk of becoming infected and developing complications [3]. Humans can be infected through the consumption of contaminated food and water, direct contact with infected animals and their environment, or via person-to-person transmission. A necessity in the successful treatment, prevention, and control of foodborne disease outbreaks is the rapid and accurate identification of the offending pathogen [4].

Bacterial subtyping of isolates of the same genus and species is an important tool in disease surveillance, outbreak investigations, and epidemiological research. A number of phenotypic and genotypic subtyping methods are available [5]; however, for foodborne pathogens, such as *Salmonella enterica* subsp. *enterica*, serotyping is often among the most widely used initial characterizations performed on isolates. *Salmonella* serotyping is a phenotypic subtyping method that has been in existence for over 80 years and is still the primary screening method in many laboratories [6]. The basis of this method of subtyping is observation for agglutination reactions occurring between specific antisera and somatic (O) and flagellar (H) antigens of the *Salmonella* isolate. The *Salmonella* isolate is then classified using the Kauffman–White–LeMinor scheme [7]. However, since there are over 2500 serovars of *Salmonella enterica*, with 46 O antigens and 114 H variations [8], serotyping can be quite exhausting and time-consuming and require a vast number of antisera [9]. Additionally, the possibility exists for inaccurate classification as a result of observer's error, nonspecific agglutination, auto-agglutination, or loss of antigenic expression [6,10].

In comparison, Matrix Laser Desorption–Ionization Time of Flight (MALDI–TOF), a library-based approach to bacterial identification, offers a rapid, reproducible method for bacterial identification with a high sensitivity and specificity and at minimal cost [11,12]. MALDI-TOF uses the mass-to-charge ratio profile of bacterial microbial proteins and peptides for bacterial identification [13]. This mass-to-charge ratio profile or mass spectral profile analysis is usually confined to the 2–20 kDA range, since the majority of peaks in this range are representative of ribosomal proteins which are less influenced by variability in cultivation conditions [14]. Bacterial isolates can be characterized at the genus and species level via the identification of a unique biomarker ion peak(s) or by matching the mass spectral profile or "fingerprint" of query bacteria with the spectral profiles of known bacterial species within the established MALDI-TOF library using pattern recognition algorithms [5,9,11,15]. Identification at the subspecies level can be more difficult due to the lack of unique ion peaks between serotypes and also due to poor differentiation of mass spectral profiles between serotypes [13]. Although MALDI-TOF use has been well validated for bacterial identification at the species level, differentiation at the subspecies level is less well described.

The purpose of this study was to determine the feasibility of using MALDI-TOF technology to differentiate six known *Salmonella* serovars recovered from a long-term fecal survival study. To achieve this, we (1) constructed and validates reference spectra for the six Salmonella serovars and (2) compared the results obtained from MALDI-TOF subtyping with those from conventional subtyping of *Salmonella* isolates recovered from experimental microcosms inoculated with these same strains of *Salmonella enterica* serovars.

2. Materials and Methods

Bacterial strains: The *Salmonella enterica* serovars used in this study were *S.* Anatum (K2669 CDC clinical isolate), *S.* Braenderup (04E61556), *S.* Javiana (ATCC® BAA-1593™), S. Montevideo (Human–tomato linked), *S.* Newport (Environmental isolate), and *S.* Typhimurium (ATCC® 700720™). The serovars and their epidemiological history were kindly provided by Dr. Michelle Danyluk, Citrus Research and Education Center, University of

Florida. The six *Salmonella* serovars were transformed to exhibit Rifampicin resistance at 80 µg/mL in order to distinguish the inoculated serovars from background microflora in feces. Rifampicin resistance was achieved by serial passage of the parent salmonella isolates in increasing concentrations (10, 20, 30, 40, 50, 60, 70, 80, 90, 100 µg/mL) of Rifampicin (Fischer Scientific, Fair lawn, NJ, USA). The final broth culture was plated on LB agar with rifampicin (80 µg/mL) and incubated overnight at 37 °C. The next day, three isolates were selected and grown overnight in BHI broth and stored at −80 °C with 30% buffered glycerol for use at a later time. Rifampicin resistance is conferred as a result of mutation in the beta subunit of RNA polymerase. Previous studies comparing the use of plasmids versus chromosomal mutations for bacterial identification in long-term survival studies indicated that bacteria with chromosomal mutations accurately represented the bacterial population, and this method was much more reliable than using plasmids [16].

Experimental microcosm: The fecal survival study was conducted at four different laboratories in California, Delaware, Florida, and Ohio as described in Topalcengiz et al. (2020) [15]. Briefly, the rifampicin-resistant serovars were cultured separately overnight at 37 °C in Buffered Peptone Water (BPW, Acumedia, East Lansing, MI, USA). Following incubation, 45 mL samples of each Salmonella broth culture was placed in 50 mL centrifuge tubes and centrifuged at 4600× *g* for 20 min. The supernatant was decanted, and the bacterial pellet was resuspended in 45 mL of 1× phosphate-buffered saline (PBS, AMRESCO, Salon, OH, USA). This 'washing' procedure was repeated two more times to ensure the removal of any nutrient content or antibiotic selective pressure. After the third washing and re-suspension of the bacterial pellet, the optical densities of the resuspended solutions were measured to attain an inoculation dose between 10^4 and 10^5 CFU/gram feces, and the feces were inoculated. The inoculated fecal microcosms were left at room temperature (22 ± 3 °C) until the day of sampling. The population of *Salmonella* in each fecal sample (cattle, deer, raccoon, wild hog, and waterfowl) was enumerated at days 1, 3, 5, 7, 14 and monthly by surface plating 1 mL dilutions (10^{-1}–10^{-6}) of each fecal sample on LB agar, Lennox (LB, Acumedia, East Lansing, MI, USA) plates containing 80 µg/mL of rifampicin and 50 µg/mL of cycloheximide (Sigma–Aldrich, St. Louis, MO, USA).Up to 10 (if present) *Salmonella* colonies recovered from each fecal sample on every sampling day were transferred from the LB agar plates to 2.0 mL centrifuge tubes containing 1.0 mL of brain heart infusion (BHI, Acumedia, East Lansing, MI, USA) and cultured overnight at 37 °C. The next day, 300 µL of buffered glycerol (VWR International, Radnor, PA, USA) was added to each tube, and the content mixed. These cultures were then stored at −80 °C until identification. The recovered *Salmonella* isolates from California, Florida, and Delaware were shipped to the Ohio Agriculture Research and Development Center (OARDC) for identification.

Construction of Reference Spectra: The reference *Salmonella* strains were streaked onto XLT-4 agar (Acumedia, East Lansing, MI, USA) plates and incubated at 37 °C for 18–24 h. Following incubation, one colony from each plate was selected and streaked onto LB agar plates supplemented with rifampicin (80 µg/mL) and incubated at 37 °C for 18–24 h. Following incubation, 15 colonies were selected from each serotype and subjected to protein extraction procedures as described by the manufacturer [16]. Briefly, each colony was transferred from the agar plate to a 1.7 mL microcentrifuge tube containing 300 µL of HPLC-grade water (EMD Chemicals–Gibbstown, NJ, USA). The contents of the tube were mixed thoroughly by vortexing for approximately 1 min. Following thorough mixing, 900 µL of HPLC grade Ethanol (ACROS Organics, Fischer Scientific, Fair Lawn, NJ, USA) was added to each tube, and the contents were mixed via vortexing for one minute. Each tube was then centrifuged (Microfuge 2, Beckman Coulter, CA, USA) at 18,000× *g* for 2 min at 4 °C. After centrifugation, the supernatant was carefully decanted, and the tubes centrifuged again at 18,000× *g* for 2 min at 4 °C. Following this second centrifugation procedure, the residual supernatant was carefully removed by pipetting, ensuring the bacterial pellet was not disturbed. The tubes containing the bacterial pellet were then left uncovered and allowed to air-dry at room temperature.

Following air-drying, 10 µL of 70% of Formic Acid (Fluka, Sigma Aldrich, St. Louis, MO, USA) was added to each tube, and the contents were agitated via vortexing for 1 min to ensure the resuspension of the bacterial pellets. The suspensions were then left to stand for 5 min, after which, 10 µL of HPLC-grade acetonitrile (Fischer Scientific, Fair lawn, NJ) was added to each tube and mixed thoroughly via vortexing. The tubes were then centrifuged at 18,000× *g* for 2 min at 4 °C.

After centrifugation, 1 µL of the supernatant was pipetted without disturbing the pellet and transferred to one well on the MSP 96 polished-steel target (Bruker Daltonics, Billerica, MA, USA). This was done in triplicate for each sample. The 1 µL aliquots of supernatant on the target were allowed to air-dry at room temperature, following which each well was overlayed with 1 µL of Bruker Matrix HCCA (10 mg/mL) solution (α-Cyano-4-hydroxycinnamic acid, Bruker Daltonics, Billerica, MA, USA), which had been reconstituted in 250 µL of organic solvent (comprising acetonitrile, trifluoracetic acid, and water). The steel target was then allowed to air-dry. The mass spectrum of each protein extract was then assessed using Bruker Microflex LT (Bruker Daltonics, Billerica, MA, USA), and a reference Mass Spectral Profile (MSP) was created as described below.

Mass Spectral Profile Creation: Prior to analyzing the mass spectra of protein extracts, each target was calibrated to ensure the accuracy of the measurements. Calibration was achieved by placing 1 µL of Bacterial Test Standard (BTS; Bruker Daltonics, Billerica, MA, USA) in one empty well of the target containing the protein extract samples. Once dried, this well was then overlayed with HCCA matrix as described above, and the auto-calibration option in the flexControl software (Bruker Daltonics, Billerica, MA, USA) was used to facilitate calibration. Once calibrated, spectra were acquired using the AutoXecute function within flexControl. Three spectra were recorded from each sample well, and there were 15 replicates of each serovar; thus, each reference spectrum was constructed using 45 spectra. The spectra were obtained using the recommended setting for bacterial species identification (linear positive mode, 20 Hz laser frequency, 20 kV acceleration voltage, 18.5 kV IS2 voltage, 250 ns extraction delay).

Following acquisition, all spectra for each serotype were then imported into flex-Analysis (Bruker Daltonics, Billerica, MA, USA). The spectra were subjected to baseline subtraction and mass spectrum smoothing for evaluation of their uniformity and detection of abnormal peaks, flatlines, or other anomalies. The identity of any discrepant spectra was recorded, and these spectra were removed from further analysis. All spectra passing the previous quality control screening were then imported into MALDI Biotyper 3.0 (Bruker Daltonics, Billerica, MA, USA), and the new MSP for each serotype was created. Additionally, the peak list was evaluated to ensure there were a minimum of 70 peaks per MSP and these peaks were in 90% or greater of the constituting spectra. All six MSPs were then added to create a new reference library.

Isolate identification using MALDI-TOF: The unknown *Salmonella* isolates recovered from the survival studies at each location were shipped to the OARDC, where they were cultured and processed for Intact Cell Mass Spectrometry (ICMS). Briefly, one loopful of the previously frozen culture broth was streaked for colony isolation onto LB agar plates supplemented with rifampicin (80 µg/mL) and incubated at 37 °C overnight. The following day, one individual colony was selected and smeared onto two separate wells of the MALDI-TOF target. The wells were then overlayed with 1 µL of HCCA matrix, allowed to air-dry at room temperature, and analyzed using Bruker Microflex LT (Bruker Daltonics, Billerica, MA, USA), with the instrument settings described earlier. Identification of the samples was done automatically using Bruker Realtime Classification software (Bruker Daltonics, Billerica, MA, USA) by comparing the isolate MSP with the 6 newly created reference MSP.

Isolate identification using Serology and PCR: The identities of the isolates were confirmed via a combination of serotyping of somatic antigens and a PCR-based assay. This was done independently of the MALDI-TOF analysis to ensure there was no perceived bias. The unknown isolates were streaked for colony isolation onto LB agar plates supplemented with rifampicin (80 μg/mL) and incubated at 37 °C overnight. The next day, one individual colony was selected from each plate and mixed with 10 μL of *Salmonella* antisera[k] for somatic groups B, C1, C2, D, and E1 and observed for agglutination. Two of the serotypes, *S.* Montevideo and *S.* Braenderup, belong to Group C1 and agglutinated equally when mixed with C1 antisera. The differentiation of these two serotypes was accomplished via a PCR-based assay adapted from J. Jean-Gilles Beaubrun et al. (2012) [17]. Briefly, individual colonies from samples agglutinating with C1 were placed in 1.7 mL microcentrifuge tubes containing 100 μL of sterile DNAase-free water. The tubes were then placed in a water bath at 100 °C for 10 min. The cell lysates were then used as DNA templates for PCR screening for the detection of *stm1350* (171 bp) and *sty0346* and *sty0347* (262 bp) [17]. The primer sequences used were *stm1350F:* 5′TCAAAATTACCGGGCGCA3′; *stm1350R:* 5′TTTTAAGACTACATACGCGCATGAA3′; *STY0346:* 5′GGCTGGAGCAGCCTTACAAAA3′; and *sty0347* 5′AAGAGTTGCCTGGCTGG TAAAA3′. Amplifications were performed in a 50 μL reaction mixture containing 25 μL GoTaq Green Mastermix (Fischer Scientific, Fair lawn, NJ, USA), 2 μL *sty0346* (5 μM/μL), 2 μL *sty0347* (5 μM/μL), 2 μL *stm1350* F (5 μM/μL), 2 μL *stm1350* R (5 μM/μL), 14 μL H_2O, and 3 μL DNA template. The reaction mixture was heated to initial denaturation at 94 °C for 5 min, followed by 35 cycles at 94 °C for 30 s, 52 °C for 1 min, 72 °C for 1 min, and final extension at 72 °C for 5 min. The PCR products were separated on 2% agarose gels stained with ethidium bromide and visualized using UV light.

Data Analysis: The results were tabulated using 2 × 2 contingency tables for each serotype to reflect the agreement between the results obtained from MALDI-TOF and those from serotyping/PCR. The overall percent agreement (overall number of isolates positively and negatively identified as a particular by both MALDI TOF and Serology/Number of isolates tested), positive percent agreement (Number Samples positively identified as a particular serotype by both MALDI-TOF and Conventional Serology/Number of Samples identified as the serotype using conventional serology), negative percent agreement (Number Samples identified as not being a particular serotype by both MALDI-TOF and Conventional Serology/Number of Samples identified as the serotype using conventional serology), and Cohen's kappa were calculated to evaluate the agreement between the methods. Cohen's Kappa provided a measure of the degree to which the two methods concurred in their respective classification of the different serovars [18,19]. Statistical calculations were done using VassarStats [20].

3. Results

Spectra Validation: To validate the accuracy of the created MSPs, 40 samples of each serotype whose identity was known were run against the newly created MSP library. All correctly identified reference isolates had an overall mean score for all serotypes of 2.54 (95% CI: 2.52–2.56). For the identification and subtyping of the isolates, we conservatively used the lower limit of 2.52 as our cut-off. If an unknown isolate achieved a biotyper score of less than 2.52 for the best match when analyzed, the isolates were restreaked, and fresh colonies were reanalyzed.

Salmonella subtyping: MALDI-TOF subtyped 95% (803/845) of *Salmonella* isolates the same as the conventional serotyping and PCR combination. The majority of isolates recovered, 430, belonged to the serovar *S.* Javiana, while the least number of recovered isolates, 17, belonged to the *S.* Newport serovar (Table 1). The positive percent agreement, which is a measure of MALDI-TOF ability to identify a serovar compared to the conventional methods, was lowest for *S.* Montevideo (79%) and highest for *S.* Typhimurium and *S.* Newport, where 100% of isolates were correctly matched. The negative percent agreement, which is a measure of MALDI-TOF ability to correctly classify a sample as not being a particular serotype, was above 98% for all serotypes.

Table 1. Comparison of the results obtain from MALDI-TOF and conventional subtyping methods.

Salmonella Serotype *	Number ** of Isolates	MALDI–TOF ***						OPA §	PPA †	NPA ‡	Cohen's Kappa #
		S.T	S.M	S.A	S.J	S.B	S.N				
S.T	38	38						98.6%	100%	98.5%	0.86
S.M	48	5	38	4	1			98.7%	79.1%	99.8%	0.87
S.A	156	5		144		2	5	96.8%	92.3%	97.8%	0.89
S.J	430		1		429			99.0%	98.8%	99.0%	0.98
S.B	154			11	3	137	1	97.7%	89.0%	99.7%	0.92
S.N	17						17	99.3%	100%	99.3%	0.85

* S.T—*Salmonella* Typhimurium, S.M—*Salmonella* Montevideo, S.A—*Salmonella* Anatum, S.J—*Salmonella* Javiana; S.B.—*Salmonella* Braenderup, S.N—*Salmonella* Newport; ** Number of isolates identified using conventional serology/PCR; *** Number of isolates identified using MALDI-TOF; I§ OPA—Overall Percent Agreement; † PPA—Positive Percent Agreement; ‡ NPA—Negative Percent Agreement; # Kappa scores greater than 0.81 are considered to indicate an almost perfect agreement between the testing methods [18,19].

We also calculated Cohen's Kappa as another measure to evaluate the agreement between the subtyping results obtained by MALDI-TOF and by the conventional serotyping/PCR method. The subtyping of *S.* Newport received the lowest Cohen's Kappa score (0.85), while the subtyping of *S.* Javiana received the highest score (0.98) (Table 1).

4. Discussion

Herein, we report that MALDI-TOF is capable of subtyping *Salmonella* using mass spectral profile analysis equally as well as conventional methods (serotyping + PCR), while at the same time allowing a rapid identification at a reduced cost. We can thus propose that MALDI-TOF can be used as an alternative method for the rapid identification of *Salmonella* serovars used in epidemiological studies such as this one.

Surprisingly, although MALDI-TOF technology has been in existence for over 20 years there is a paucity of published literature describing its ability and use to identify *Salmonella* serovars. At the species level, *Salmonella enterica* subspecies enterica was first differentiated from other *Salmonella enterica* subspecies using mass spectrometry in 2008 [21]. Serovar differentiation capability was also first reported by Dieckmann et al. in 2011, when he reported successful ICMS differentiation of *S.* Enteriditis, *S.* Typhimurium, *S.* Virchow *S.* Infantis, and *S.* Hadar [9]. The authors further postulated the existence of potential discriminatory biomarker ions in *Salmonella enterica* serotypes, but these have not been extensively evaluated. Discrimination at this phylogenic level was later demonstrated by Kuhns et al., (2012), who were able to differentiate *Salmonella* Typhi from other serovars, but not to differentiate between the other 11 serovars of *Salmonella enterica* subspecies *enterica* tested [22]. Discrimination between bacterial serovar strains may be required in disease outbreaks, especially foodborne disease outbreaks. While this strain discrimination is possible for certain bacterial species, this has not been the case for *Salmonella* species using conventional MALDI-TOF [23]. Additional research is required to evaluate to potential for MALDI-TOF MS combined with principal component analysis or MALDI-TOF MS–MS to provide this level of discriminatory power, especially if MALDI-TOF is to be used in foodborne disease outbreak investigations [13,24]. With conventional MALDI-TOF analysis the reproducibility of unique strain specific biomarker peaks necessary for *Salmonella* spp strain identification has historically posed a problem, but recent studies have postulated the reproducibility of these peaks may be increased by controlling bacterial concentration using the supernatant obtained after centrifuging the colony suspension, and including various matrix additives [25]. The potential to identify both serovar- and strain-specific biomarker peaks will undoubtedly increase the usefulness and precision of MALDI-TOF in clinical and epidemiological investigations.

MALDI-TOF biotyper scores are assigned based on the similarity of an organism's mass spectrum to the reference spectra [14]. The quality of the spectra can be dependent on a number of factors including age of culture, sample preparation, thickness of colony smear on target, and matrix used [26–28]. The difference between the methods used for the creation of the reference MSP and the analysis of unknown isolates, i.e., protein extraction vs. intact cell mass spectrometry (ICMS), can potentially affect the identification of organisms [29]. This effect is more pronounced with Gram-positive organisms and yeast cells, whose the thicker cell wall may affect identification when the direct smear method is used [30,31]. For Gram-negative bacteria, such as *Salmonella* spp., this is less pronounced and may only lead to discordant results in a small number of isolates; consequently, the direct smear approach for the analysis of these isolates, which is faster and more economical, is suitable [30,32]. However, in cases where strain identification is necessary, for example in foodborne disease outbreaks, it may be necessary to use the protein extraction technique to enhance the expression of biomarker peaks in the analyte and thus improve the accuracy of the identification. Ford et al. (2013) also reported that rate of identification dropped as the culture age increased, and Veelo et al., (2014) further reported that thickness of the smear on the target can affect the identification of organisms [27,33]. It is for these reasons, if isolates obtained low scores below our cut-off values, that we repeated the culture and smear preparation one more time and reanalyzed the isolates with reference strains to validate the preparation procedures.

Biotyper scores are logarithmic, ranging between 0 and 3, and high values in this score represent a high similarity with the database entry [34]. Currently, MALDI-TOF scores greater than 2.3 are suggested by the manufacturers to indicate a high probability of species identification. However, for this experiment, a higher discriminatory power for serovar identification is required, and no MALDI-TOF scores are suggested for this level of discrimination. Consequently, since we required an increased similarity match for serovars versus that normally required for species identification, we increased our serovar classification score above the value of 2.3 suggested for species identification and used a score based on screening of the reference spectra against the newly created *Salmonella* library as our cut-off. Adjustment to biotyper cut-off scores is not unique, and other authors have also proposed adjustments to biotyper classification cut-off scores to optimize the classification of organisms, but these studies were confined to improve genus and species classification [35–37] and not serovar classification, as we showed.

Although the initial cost of MALDI-TOF machines can be high, the money saved from the cost of diagnostic reagents and decreased labor requirements, coupled with the rapid identification time, makes MALDI-TOF identification of bacteria a suitable alternative to conventional organism identification methods [14]. These benefits are most pronounced in epidemiological studies like this one, where large numbers of samples are processed. For example, the average cost of consumables used for serotyping a *Salmonella* isolate is estimated to be USD 8.00 [38], while consumables used in MALDI-TOF costs an estimated USD 0.50 [14,39]. The difference in processing time was also very apparent, since the identification of a *Salmonella* isolates using the combination of serotyping and PCR took at most 3 h for some isolates, while MALDI-TOF identification took as little as 3 min per isolate, including the time necessary for sample preparation and MALDI-TOF analysis.

In conclusion, the MALDI-TOF method proposed here as a rapid, cost-effective method for the identification of *Salmonella* serovars was proven to have equal diagnostic capabilities as conventional subtyping methods. Consequently, this method can be used to complement conventional methods of serovar identification; however, if definitive serovar identification and strain discrimination are required, as in the case of outbreak scenarios, the identification of isolates will still rely on molecular confirmatory tests.

Author Contributions: Conceptualization, R.L.B., M.T.J.-R., K.E.K., M.D.D. and J.T.L.; Formal analysis, A.K.P.; Funding acquisition, R.L.B., M.T.J.-R., K.E.K., M.D.D. and J.T.L.; Investigation, A.K.P., H.A.F., N.A., J.C., Z.T., S.J., P.M.S., G.R. and J.T.L.; Methodology, A.K.P., H.A.F., N.A., Z.T., S.J., P.M.S., G.R. and J.T.L.; Supervision, J.T.L.; Writing—original draft, A.K.P.; Writing—review & editing, A.K.P., J.C., Z.T., S.J., P.M.S., R.L.B., M.T.J.-R., K.E.K., M.D.D., G.R. and J.T.L. All authors have read and agreed to the published version of the manuscript.

Funding: This material is based upon work that is supported by the National Institute of Food and Agriculture, U.S. Department of Agriculture, Specialty Crops Research Initiative (SCRI), under awards SCRI-2011-51181-30767. Any opinions, findings, conclusions, or recommendations expressed in this publication are those of the author(s) and do not necessarily reflect the view of the U.S. Department of Agriculture.

Institutional Review Board Statement: Not applicable.

Informed Consent Statement: Not applicable.

Data Availability Statement: The data that support the findings of this study are available on request from the corresponding author. The data are not publicly available due to privacy restrictions.

Acknowledgments: The authors would also like to thank Michael Kauffman, Jennifer Schrock, Dominika Jurkovic, and Mary-Beth Weisner for their assistance in conducting this study.

Conflicts of Interest: The authors declared no potential conflict of interest with respect to the research, authorship, and/or publication of this article.

References

1. Lee, K.-M.; Runyon, M.; Herrman, T.J.; Phillips, R.; Hsieh, J. Review of Salmonella detection and identification methods: Aspects of rapid emergency response and food safety. *Food Control* **2015**, *47*, 264–276. [CrossRef]
2. Scallan, E.; Hoekstra, R.M.; Angulo, F.J.; Tauxe, R.V.; Widdowson, M.A.; Roy, S.L.; Jones, J.L.; Griffin, P.M. Foodborne illness acquired in the United States-major pathogens. *Emerg. Infect. Dis.* **2011**, *17*, 7–15. [CrossRef] [PubMed]
3. Baird-Parker, A.C. Originally published as Volume 336, Issue 8725Foodborne salmonellosis. *Lancet* **1990**, *336*, 1231–1235. [CrossRef]
4. Sabat, A.; Budimir, A.; Nashev, D.; Sá-Leão, R.; Van Dijl, J.; Laurent, F.; Grundmann, H.; Friedrich, A.; Markers, E.S.G.o.E. Overview of molecular typing methods for outbreak detection and epidemiological surveillance. *Euro Surveill. Bull. Eur. Sur Les Mal. Transm.= Eur. Commun. Dis. Bull.* **2013**, *18*, 20380. [CrossRef]
5. Carbonnelle, E.; Mesquita, C.; Bille, E.; Day, N.; Dauphin, B.; Beretti, J.L.; Ferroni, A.; Gutmann, L.; Nassif, X. MALDI-TOF mass spectrometry tools for bacterial identification in clinical microbiology laboratory. *Clin. Biochem.* **2011**, *44*, 104–109. [CrossRef]
6. Wattiau, P.; Boland, C.; Bertrand, S. Methodologies for Salmonella enterica subsp. enterica subtyping: Gold standards and alternatives. *Appl. Environ. Microbiol.* **2011**, *77*, 7877–7885. [CrossRef]
7. Available online: https://www.academia.edu/download/41403599/Antigenic_Formulae_of_the_Salmonella_ser20160122-19332-1wjmz5c.pdf (accessed on 1 March 2016).
8. McQuiston, J.R.; Fields, P.I.; Tauxe, R.V.; Logsdon, J.M., Jr. Do Salmonella carry spare tyres? *Trends Microbiol.* **2008**, *16*, 142–148. [CrossRef]
9. Dieckmann, R.; Malorny, B. Rapid screening of epidemiologically important Salmonella enterica subsp. enterica serovars by whole-cell matrix-assisted laser desorption ionization-time of flight mass spectrometry. *Appl. Environ. Microbiol.* **2011**, *77*, 4136–4146. [CrossRef]
10. Schrader, K.N.; Fernandez-Castro, A.; Cheung, W.K.; Crandall, C.M.; Abbott, S.L. Evaluation of commercial antisera for Salmonella serotyping. *J. Clin. Microbiol.* **2008**, *46*, 685–688. [CrossRef]
11. Kliem, M.; Sauer, S. The essence on mass spectrometry based microbial diagnostics. *Curr. Opin. Microbiol.* **2012**, *15*, 397–402. [CrossRef]
12. Christner, M.; Trusch, M.; Rohde, H.; Kwiatkowski, M.; Schluter, H.; Wolters, M.; Aepfelbacher, M.; Hentschke, M. Rapid MALDI-TOF mass spectrometry strain typing during a large outbreak of Shiga-Toxigenic Escherichia coli. *PLoS ONE* **2014**, *9*, e101924. [CrossRef]
13. Sandrin, T.R.; Goldstein, J.E.; Schumaker, S. MALDI TOF MS profiling of bacteria at the strain level: A review. *Mass Spectrom. Rev.* **2013**, *32*, 188–217. [CrossRef]
14. Dingle, T.C.; Butler-Wu, S.M. Maldi-tof mass spectrometry for microorganism identification. *Clin. Lab. Med.* **2013**, *33*, 589–609. [CrossRef]
15. Topalcengiz, Z.; Spanninger, P.M.; Jeamsripong, S.; Persad, A.K.; Buchanan, R.L.; Saha, J.; LeJEUNE, J.; Jay-Russell, M.T.; Kniel, K.E.; Danyluk, M.D. Survival of Salmonella in various wild animal feces that may contaminate produce. *J. Food Prot.* **2020**, *83*, 651–660. [CrossRef]
16. Schumann, P.; Maier, T. MALDI-TOF mass spectrometry applied to classification and identification of bacteria. *Methods Microbiol.* **2014**, *41*, 275–306.
17. Jean-Gilles Beaubrun, J.; Cheng, C.M.; Chen, K.S.; Ewing, L.; Wang, H.; Agpaoa, M.C.; Huang, M.C.; Dickey, E.; Du, J.M.; Williams-Hill, D.M.; et al. The evaluation of a PCR-based method for identification of Salmonella enterica serotypes from environmental samples and various food matrices. *Food Microbiol.* **2012**, *31*, 199–209. [CrossRef]
18. Landis, J.R.; Koch, G.G. An application of hierarchical kappa-type statistics in the assessment of majority agreement among multiple observers. *Biometrics* **1977**, *33*, 363–374. [CrossRef]

19. Landis, J.R.; Koch, G.G. The measurement of observer agreement for categorical data. *biometrics* **1977**, *33*, 159–174. [CrossRef]
20. Available online: http://vassarstats.net/ (accessed on 7 December 2021).
21. Dieckmann, R.; Helmuth, R.; Erhard, M.; Malorny, B. Rapid classification and identification of salmonellae at the species and subspecies levels by whole-cell matrix-assisted laser desorption ionization-time of flight mass spectrometry. *Appl. Environ. Microbiol.* **2008**, *74*, 7767–7778. [CrossRef]
22. Kuhns, M.; Zautner, A.E.; Rabsch, W.; Zimmermann, O.; Weig, M.; Bader, O.; Gross, U. Rapid discrimination of Salmonella enterica serovar Typhi from other serovars by MALDI-TOF mass spectrometry. *PLoS ONE* **2012**, *7*, e40004. [CrossRef]
23. Kang, L.; Li, N.; Li, P.; Zhou, Y.; Gao, S.; Gao, H.; Xin, W.; Wang, J. MALDI-TOF mass spectrometry provides high accuracy in identification of Salmonella at species level but is limited to type or subtype Salmonella serovars. *Eur. J. Mass Spectrom.* **2017**, *23*, 70–82. [CrossRef]
24. Yan, W.; Qian, J.; Ge, Y.; Ye, K.; Zhou, C.; Zhang, H. Principal component analysis of MALDI-TOF MS of whole-cell foodborne pathogenic bacteria. *Anal. Biochem.* **2020**, *592*, 113582. [CrossRef] [PubMed]
25. Fukuyama, Y.; Ojima-Kato, T.; Nagai, S.; Shima, K.; Funatsu, S.; Yamada, Y.; Tamura, H.; Nomura, S.; Ogata, K.; Sekiya, S. Improved MALDI-MS method for the highly sensitive and reproducible detection of biomarker peaks for the proteotyping of Salmonella serotypes. *J. Mass Spectrom.* **2019**, *54*, 966–975. [CrossRef] [PubMed]
26. Giebel, R.; Worden, C.; Rust, S.M.; Kleinheinz, G.T.; Robbins, M.; Sandrin, T.R. Microbial fingerprinting using matrix-assisted laser desorption ionization time-of-flight mass spectrometry (MALDI-TOF MS) applications and challenges. *Adv. Appl. Microbiol.* **2010**, *71*, 149–184. [CrossRef] [PubMed]
27. Ford, B.A.; Burnham, C.A. Optimization of routine identification of clinically relevant Gram-negative bacteria by use of matrix-assisted laser desorption ionization-time of flight mass spectrometry and the Bruker Biotyper. *J. Clin. Microbiol.* **2013**, *51*, 1412–1420. [CrossRef] [PubMed]
28. Holland, R.D.; Wilkes, J.G.; Cooper, W.M.; Alusta, P.; Williams, A.; Pearce, B.; Beaudoin, M.; Buzatu, D. Thymol treatment of bacteria prior to matrix-assisted laser desorption/ionization time-of-flight mass spectrometric analysis aids in identifying certain bacteria at the subspecies level. *Rapid Commun. Mass Spectrom RCM* **2014**, *28*, 2617–2626. [CrossRef] [PubMed]
29. Cuénod, A.; Foucault, F.; Pflüger, V.; Egli, A. Factors associated with MALDI-TOF mass spectral quality of species identification in clinical routine diagnostics. *Front. Cell. Infect. Microbiol.* **2021**, *11*, 104. [CrossRef] [PubMed]
30. Bizzini, A.; Durussel, C.; Bille, J.; Greub, G.; Prod'Hom, G. Performance of matrix-assisted laser desorption ionization-time of flight mass spectrometry for identification of bacterial strains routinely isolated in a clinical microbiology laboratory. *J. Clin. Microbiol.* **2010**, *48*, 1549–1554. [CrossRef]
31. Fournier, R.; Wallet, F.; Grandbastien, B.; Dubreuil, L.; Courcol, R.; Neut, C.; Dessein, R. Chemical extraction versus direct smear for MALDI-TOF mass spectrometry identification of anaerobic bacteria. *Anaerobe* **2012**, *18*, 294–297. [CrossRef]
32. Mestas, J.; Quias, T.; Dien Bard, J. Direct Identification of Aerobic Bacteria by Matrix-Assisted Laser Desorption Ionization Time-of-Flight Mass Spectrometry Is Accurate and Robust. *J. Clin. Lab. Anal.* **2016**, *30*, 543–551. [CrossRef]
33. Veloo, A.C.; Elgersma, P.E.; Friedrich, A.W.; Nagy, E.; van Winkelhoff, A.J. The influence of incubation time, sample preparation and exposure to oxygen on the quality of the MALDI-TOF MS spectrum of anaerobic bacteria. *Clin. Microbiol. Infect. Off. Publ. Eur. Soc. Clin. Microbiol. Infect. Dis.* **2014**, *20*, O1091–O1097. [CrossRef]
34. Richter, C.; Hollstein, S.; Woloszyn, J.; Kaase, M.; Gatermann, S.G.; Szabados, F. Evaluation of species-specific score cut-off values for various Staphylococcus species using a MALDI Biotyper-based identification. *J. Med. Microbiol.* **2012**, *61*, 1409–1416. [CrossRef]
35. Schmitt, B.H.; Cunningham, S.A.; Dailey, A.L.; Gustafson, D.R.; Patel, R. Identification of anaerobic bacteria by Bruker Biotyper matrix assisted laser desorption ionization-time of flight mass spectrometry with on-plate formic acid preparation. *J. Clin. Microbiol.* **2013**, *5*, 782–786. [CrossRef]
36. Schulthess, B.; Bloemberg, G.V.; Zbinden, R.; Böttger, E.C.; Hombach, M. Evaluation of the Bruker MALDI Biotyper for Identification of Gram-Positive Rods: Development of a Diagnostic Algorithm for the Clinical Laboratory. *J. Clin. Microbiol.* **2014**, *52*, 1089–1097. [CrossRef]
37. Barberis, C.; Almuzara, M.; Join-Lambert, O.; Ramírez, M.S.; Famiglietti, A.; Vay, C. Comparison of the Bruker MALDI-TOF Mass Spectrometry System and Conventional Phenotypic Methods for Identification of Gram-Positive Rods. *PLoS ONE* **2014**, *9*, e106303. [CrossRef]
38. Association of Public Health Laboratories. Salmonella Serotyping in US Public Health Laboratories. 2014. Available online: http://www.aphl.org/AboutAPHL/publications/Documents/FS_SalmonellaSustainabilityWhitePaper_Nov2014.pdf (accessed on 1 February 2022).
39. Singhal, N.; Kumar, M.; Kanaujia, P.K.; Virdi, J.S. MALDI-TOF mass spectrometry: An emerging technology for microbial identification and diagnosis. *Front. Microbiol.* **2015**, *6*, 791. [CrossRef]

microorganisms

MDPI

Article

Molecular Characterization of *Salmonella* Detected along the Broiler Production Chain in Trinidad and Tobago

Anisa Sarah Khan [1], Rian Ewald Pierneef [2], Narjol Gonzalez-Escalona [3], Meghan Maguire [3], Cong Li [4], Gregory H. Tyson [4], Sherry Ayers [4], Karla Georges [1], Woubit Abebe [5] and Abiodun Adewale Adesiyun [1,6,*]

1 School of Veterinary Medicine, Faculty of Medical Sciences, University of the West Indies, St. Augustine 999183, Trinidad and Tobago; anisakhan11@gmail.com (A.S.K.); karla.georges@sta.uwi.edu (K.G.)
2 Agricultural Research Council-Biotechnology Platform, Onderstepoort, 100 Old Soutpan Road, Pretoria 0110, South Africa; pierneefr@arc.agric.za
3 Division of Microbiology, Office of Regulatory Science, Center for Food Safety and Applied Nutrition, Food and Drug Administration, College Park, MD 20740, USA; narjol.gonzalez-escalona@fda.hhs.gov (N.G.-E.); meghan.maguirethon@fda.hhs.gov (M.M.)
4 Division of Animal and Food Microbiology, Office of Research, Center for Veterinary Medicine, U.S. Food and Drug Administration, Laurel, MD 20708, USA; cong.li@fda.hhs.gov (C.L.); gregory.tyson@fda.hhs.gov (G.H.T.); sherry.ayers@fda.hhs.gov (S.A.)
5 Department of Pathobiology, College of Veterinary Medicine, Tuskegee University, 201 Frederick D Patterson Dr, Tuskegee, AL 36088, USA; wabdela@tuskegee.edu
6 Department of Production Animal Studies, Faculty of Veterinary Science, Onderstepoort, University of Pretoria, Private Bag X04, Pretoria 0110, South Africa
* Correspondence: abiodun.adesiyun@sta.uwi.edu

Abstract: This cross-sectional study determined the serovars, antimicrobial resistance genes, and virulence factors of *Salmonella* isolated from hatcheries, broiler farms, processing plants, and retail outlets in Trinidad and Tobago. *Salmonella* in silico serotyping detected 23 different serovars where Kentucky 20.5% (30/146), Javiana 19.2% (28/146), Infantis 13.7% (20/146), and Albany 8.9% (13/146) were the predominant serovars. There was a 76.0% (111/146) agreement between serotyping results using traditional conventional methods and whole-genome sequencing (WGS) in in silico analysis. In silico identification of antimicrobial resistance genes conferring resistance to aminoglycosides, cephalosporins, peptides, sulfonamides, and antiseptics were detected. Multidrug resistance (MDR) was detected in 6.8% (10/146) of the isolates of which 100% originated from broiler farms. Overall, virulence factors associated with secretion systems and fimbrial adherence determinants accounted for 69.3% (3091/4463), and 29.2% (1302/4463) counts, respectively. Ten of 20 isolates of serovar Infantis (50.0%) showed MDR and contained the $bla_{CTX\text{-}M\text{-}65}$ gene. This is the first molecular characterization of *Salmonella* isolates detected along the entire broiler production continuum in the Caribbean region using WGS. The availability of these genomes will help future source tracking during epidemiological investigations associated with *Salmonella* foodborne outbreaks in the region and worldwide.

Keywords: broiler production chain; *Salmonella*; molecular characterization; whole-genome sequencing; virulence genes; antimicrobial resistance genes; *blaCTX-M-65*; Trinidad and Tobago

Citation: Khan, A.S.; Pierneef, R.E.; Gonzalez-Escalona, N.; Maguire, M.; Li, C.; Tyson, G.H.; Ayers, S.; Georges, K.; Abebe, W.; Adesiyun, A.A. Molecular Characterization of *Salmonella* Detected along the Broiler Production Chain in Trinidad and Tobago. *Microorganisms* **2022**, *10*, 570. https://doi.org/10.3390/microorganisms10030570

Academic Editor: Lisa Gorski

Received: 10 February 2022
Accepted: 4 March 2022
Published: 6 March 2022

1. Introduction

Since the 1950s, *Salmonella* has been highlighted as an economically important zoonotic pathogen by the World Health Organization (WHO) and the Food and Agriculture Organization of the United Nations (FAO) [1]. The ability of *Salmonella* to cause self-limiting gastroenteritis, coupled with high mortality rates in humans due to invasive infections are causes for public health concerns [2,3]. While many animals serve as reservoirs for *Salmonella*, poultry, and poultry products are one of the primary sources of salmonellosis in humans. Therefore, the possibility of transmission from reservoirs to other animals and humans is concerning. This is compounded by antimicrobial-resistant *Salmonella* strains within the environment, necessitating surveillance and control measures among suspected reservoirs such as chickens. In addition, *S.* Typhimurium and *S.* Enteritidis are of public health significance due to their ability to cause disease in humans and animals in developed and developing countries. However, variations in *Salmonella* serovar distribution have been reported in different countries and are said to be a function of geographic location [4,5].

The use of antimicrobial agents in food-producing animals has been implicated in developing multidrug-resistant (MDR) microorganisms and spreading them through the food chain [6,7]. Of importance to human health, some cephalosporins (β-lactams), quinolones, and aminoglycosides have been classified by the World Health Organization (WHO) as critically important agents since they are used in the treatment of extra-intestinal salmonellosis [8]. The use of ciprofloxacin and ceftiofur as the established therapy protocol for human salmonellosis could be jeopardized as genetic mechanisms promoting MDR isolates have been reported [9]. β-lactamases constitute the primary mechanism of cephalosporin resistance via enzymatic modification, where different genes are implicated. The extended-spectrum β-lactamases (ESβL) include certain alleles of bla_{TEM}, and all alleles of $bla_{\mathrm{CTX\text{-}M}}$ and bla_{SHV} genes. Extended-spectrum cephalosporins can also be hydrolyzed by the AmpC β-lactamases, of which bla_{CMY} is the most common of particular importance. Quinolone resistance was initially known to develop through chromosomal mutations [10]. However, the recent emergence of plasmid-mediated quinolone resistance (PMQR) mechanisms has been reported. These include *qnr* genes: *qnrA*, *qnrB*, *qnrS*, *qnrC*, and *qnrD*, that encode pentapeptide repeat proteins that bind to and protect DNA topoisomerase IV from inhibition by quinolones, the *aac (6′)-Ib-cr* (modified acetyltransferase) and *qepA* (efflux pump) genes, respectively [11,12]. Plasmids are traditionally known to carry antimicrobial-resistant genes and several virulence-associated traits; however, other resistance mechanisms have been reported in *Salmonella* elsewhere [13,14]. Increasing trends of resistance to quinolones and 3rd generation cephalosporins such as ciprofloxacin and ceftiofur in clinical isolates have led to the introduction of carbapenems and colistin as critical antibiotics of last resort in human salmonellosis [8]. However, the use of colistin to treat both humans and animals has resulted in the emergence of mobilized colistin resistance (*mcr*) genes [15,16]. To date, nine variants of *mcr* genes have been detected in *Salmonella* isolated from humans and animals [16–18].

Similar genetic determinants conferring resistance to aminoglycosides, tetracyclines, beta-lactams, and fluoroquinolones have been detected in *Salmonella* strains isolated from livestock and humans, concluding that food and environmental contamination from livestock are carriers of antimicrobial-resistant (AMR) *Salmonella* and are sources of infection to humans [19–21]. Thus, it is critical to investigate the resistance profiles and phenotypes they exhibit, and the mutations responsible for resistance using molecular analysis methods.

Therefore, the objectives of this study were to determine the genotypic profiles (serovar, antimicrobial resistance, and virulence factors) of *Salmonella* isolated from various stages of the broiler production–processing–retailing chain in Trinidad and Tobago.

2. Materials and Methods

2.1. Sample Selection

A total of 146 isolates of *Salmonella* used in this study originated from prior studies conducted at hatcheries and broiler farms [22], broiler processing plants [23], and retail outlets (pluck shops and supermarkets) [24,25]. The type of samples collected from the various studies are as follows, hatcheries: broken eggshells, eggs in the hatcher, eggs in the incubator, hatcher environmental swabs, hatcher fluff, and stillborn chicks; broiler farms: boot swabs, cloacal swabs, litter drag swabs, feed, and water samples—in-house supply and storage tank; processing plants: chilled chicken parts, chilled whole carcasses, neck skins, pre-evisceration carcasses, and post-evisceration carcasses; retail outlets: chicken carcasses. From a total of 207 duplicate isolates (from different enrichment and selective media) of *Salmonella*, which represented 23 serovars from the aforementioned sources, the selected

146 isolates were representatives of the serovars recovered from all *Salmonella*-positive samples. Briefly, samples were processed to isolate *Salmonella* using two enrichments broths, Rappaport-Vassiliadis Soya (RVS) and tetrathionate (TT) (Oxoid, Hampshire, England), and two selective agar, brilliant green agar (BGA) and xylose lysine tergitol 4 (XLT-4) selective media (Oxoid, Hampshire, England) [26]. Suspected *Salmonella* colonies (pink isolated colonies on BGA, red colonies with black centers on XLT-4) were subjected to biochemical tests for identification of *Salmonella* spp. using standard methods [27]. Isolates of *Salmonella* recovered from the four combinations of media (RVS/BGA, RVS/XLT-4, TT/BGA, and TT/XLT-4) were initially screened using the conventional slide agglutination test. Thereafter, 146 non-duplicate isolates of *Salmonella,* randomly selected to represent the serovars and positive samples were subjected to whole-genome sequencing. The following is a summary of the number of isolates included from earlier studies: hatcheries (n = 10), farms (n = 20), processing plant (n = 61), and retail outlets (n = 55). Five additional human clinical isolates of *Salmonella* obtained from the Caribbean Public Health Agency (CARPHA) were included in our panel of isolates subjected to WGS.

2.2. DNA Extraction and Sequencing

DNA was extracted using the Maxwell RSC cultured cells DNA kit with a Maxwell RSC instrument (Promega Corporation, Madison, WI, USA) following the manufacturer's protocols for Gram-negative bacteria with additional RNase treatment. DNA concentrations were measured with a Qubit fluorometer (Life Technologies, Carlsbad, CA, USA), standardized to 0.2 ng/µL, and the samples were stored at 4 °C before library preparation.

Whole-genome sequencing (WGS) of the *Salmonella* isolates was performed by the Public Health Agency of Canada (PHAC) Laboratory and Food and Drug Administration (FDA): Center for Food Safety and Applied Nutrition genomics laboratory (FDA-CFSAN) and Center for Veterinary Medicine (FDA-CVM), Maryland, USA. The WGS data was generated on an Illumina MiSeq using 2× 250 bp and 2 × 300 bp paired-end chemistry (Illumina Inc., San Diego, CA, USA) according to the manufacturer's instructions, at 50–150X coverage. According to the manufacturer's instructions, the libraries were constructed using 100 ng of genomic DNA using the Illumina DNA Prep (M) Tagmentation kit (Illumina Inc., San Diego, CA, USA) and the Nextera XT kit (Illumina Inc., San Diego, CA, USA).

2.3. Genomic Data Analysis and In Silico Determination of Genetic Elements

Quality control including adapter removal of the raw data was conducted using BBDuk (v.37.90; https://jgi.doe.gov/data-and-tools/bbtools/bb-tools-user-guide/bbduk-guide/, accessed on 23 July 2021); sourceforge.net/projects/bbmap/). SPAdes v.3.12.0 [28] was used to create a de novo assembly of each isolate. Only contigs larger than 500 bp were retained for further analysis. Serovar prediction was made using command-line version of SISTR [29] (Version: sistr_cmd v.1.1.1).

Gene finding in each isolate was performed with Prodigal v.2.6.3 [30] (parameters -c -n).

VFDB [31] was used to assign virulence factors. This was carried out with the predicted genes (amino acid format) from Prodigal using NCBI-blast-2.9.0+. Results were filtered for the top hit with 100% identity and 100% alignment length.

CARD [32] was used to assign antimicrobial resistance. This was performed with the predicted genes (amino acid format) from Prodigal using NCBI-blast-2.9.0+. Results were filtered for the top hit with 100% identity and 100% alignment length.

2.4. Phenotypic Methods Used for Comparison with WGS

Conventional serotyping methods using the phase reversal technique described previously [22,23,25] were performed at the Public Health Laboratory, Ministry of Health, St. Michael, Barbados. Antimicrobial resistance determined by the disk diffusion method [22–24] described previously was also used. Data generated from these two methods were compared to the genomic data.

2.5. Statistical Analyzes

R version 4.0.2 was used for Chi-square analysis and data visualization.

2.6. Data Deposition

The draft genome sequence of all *S. enterica* strains have been deposited at GenBank under the accession listed in Table S1—Metadata of 146 *Salmonella* isolates detected along the broiler production chain in Trinidad and Tobago.

3. Results

3.1. Serotyping Results

Overall, the 146 isolates of *Salmonella* subjected to conventional serotyping methods were classified into 23 serovars and 3 unspecific groups (unknown serotype). In silico analysis of the WGS data generated from these 146 isolates using the SISTR software identified 23 different serovars where Kentucky 20.5% (30/146), Javiana 19.2% (28/146), Infantis 13.7% (20/146), and Albany 8.9% (13/146) were the predominant serovars. There was a 76.0% (111/146) agreement in the test results of both methods. Isolates classified as S. Albany (n = 2), Gaminara (n = 2), Oranienburg (n = 1), and Soerenga (n = 1) by SISTR were all classified as *S.* Infantis (n = 6) using the traditional method. Three *S.* Warragul isolates detected using the conventional method were classified as *S.* Caracas on SISTR analysis. The distribution of serovars of *Salmonella* isolates from various sources is shown in Table 1.

Table 1. The distribution of serovars of *Salmonella* isolates from various sources based on in silico analysis.

	No. of Strains of *Salmonella* Detected from the Following:				
Serovars	**Hatchery**	**Farm**	**Processing Plant**	**Pluck Shop** [a]	**Supermarket** [a]
Aberdeen	0	0	1	1	0
Alachua	0	0	1	0	0
Albany	0	4	8	1	0
Anatomy	0	0	5	0	0
Caracas	0	0	0	3	0
Chester	0	0	0	0	2
Enteritidis	0	0	9	0	0
Fresno	1	0	0	0	0
Gaminara	0	3	0	0	0
Infantis	0	11	9	0	0
Javiana	0	0	10	17	1
Kentucky	8	0	7	12	3
Liverpool	0	0	1	0	0
Manhattan	0	0	0	7	0
Mbandaka	0	0	1	0	0
Molade	0	0	0	0	1
Montevideo	0	0	0	2	1
Oranienburg	0	1	0	0	0
Schwarzengrund	0	0	7	1	0
Senftenberg	1	0	0	2	1
Soerenga	0	1	0	0	0
Virchow	0	0	1	0	0
Weltevreden	0	0	1	0	0
Sub-total	10	20	61	46	9

[a] Retail outlets comprised pluck shops and supermarkets.

3.2. Antimicrobial Resistance Profiles

A total of 71 ARO accessions (Antibiotic Resistance Ontology, as defined by CARD) were detected among 22 isolates. Genes associated with aminoglycoside resistance, i.e. *aac(3)-IV* (plasmid-encoded), *aac(6′)-Iaa* (chromosomal- encoded), *aac(6′)Iy* (chromosomal encoded), *aph(3′)-Ia* (plasmid-encoded), and *aph(4)-Ia* (plasmid-encoded) (Table 2) were found at frequencies ranging from 1.4% to 7.5%. All our *S.* Manhattan and *S.* Aberdeen strains containing the often silent, chromosomal-encoded *aac(6′)-Iaa* and *aac(6′)Iy* genes, exhibited phenotypic aminoglycoside resistance. Ten (6.8%) of 146 isolates contained the *bla*$_{CTX-M-65}$ gene, which confers cephalosporin resistance. This gene was identified in *S.* Infantis isolates only. Genes *qacEDelta1* and *sul1*, responsible for antiseptic and sulfonamide resistance, were each detected at a frequency of 8.2% (12/146). *mcr-9*, the mobilized and plasmid-mediated colistin resistance gene, was found in only one isolate. Table 3 shows the distribution of AROs among *Salmonella* isolates from various sources. Isolates from broiler farms accounted for 83.1% (59/71) of AROs where the predominance of *aac(3)-IV* (9.9%; 7/71), *aph(4)-Ia* (9.9%; 7/71), *qacEdelta1* (9.9%; 7/71), *sul1* (9.9%; 7/71), and *bla*$_{CTX-M-65}$ (9.9%; 7/71) among cloacal swab isolates (62.7%; 37/59) was evident. *Salmonella* isolated from the water supply at farms (18.6%; 11/59) were found to contain 66.7% (6/9) of the AROs found in this study except for *mcr-9.1*, *aac(6′)-Iaa* and *aac(6′)-Iy*.

Table 2. Antimicrobial class and genes detected in 146 *Salmonella* isolates were used in this study.

	Antimicrobial Class and Genes Detected [a]						
Pattern	**Aminoglycoside**	**Disinfectant**	**Cephalosporin**	**Peptide**	**Sulphonamide**	**Number of Isolates (%)**	**Serovar (*n*, %)**
Pattern 1	*aph(4)-Ia* *aac(3)-IV*	*qacEDelta1*	*bla*$_{CTX-M-65}$	-	*sul1*	6 (4.2)	Infantis (6, 100.0)
Pattern 2	*aph(3′)-Ia* *aph(4)-Ia* *aac(3)-IV*	*qacEDelta1*	-	-	*sul1*	1 (0.7)	Infantis (1, 100.0)
Pattern 3	*aph(3′)Ia* *aph(4)-Ia* *aac(3)-IV*	*qacEDelta1*	*bla*$_{CTX-M-65}$	-	*sul1*	4 (2.8)	Infantis (4, 100.0)
Pattern 4	-	*qacEDelta1*	-	-	*sul1*	1 (0.7)	Senftenberg (1, 100.0)
Pattern 5	*aac(6′)-Iaa*	-	-	-	-	7 (4.9)	Manhattan (7, 100.0)
Pattern 6	*aac(6′)-Iy*	-	-	-	-	2 (1.4)	Aberdeen (2, 100.0)
Pattern 7	-	-	-	*mcr-9.1*	-	1 (0.7)	Senftenberg (1, 100.0)
Total	20 (14.9)	12 (9.0)	10 (7.5)	1 (0.7)	12 (9.0)	22 (16.1)	

[a] Of a total of 146 isolates subjected to CARD analyzes, AMR genes were detected in 22 isolates shown, 121 isolates possessing the core gene *golS* (regulator of a multidrug efflux pump) were not included in the table and three isolates were negative for resistance genes (Liverpool, Mbandaka, and Oranienburg).

Table 3. Frequency of ARO accessions detected in this study.

			Distribution of AROs among the Various Sampling Levels			
ARO Name [a]	**No. of AROs**	**Overall Frequency (%) [b]**	**Hatchery**	**Farm**	**Processing Plant**	**Retail Outlet**
aac(3)-IV	11	7.5	0	11	0	0
aac(6′)-Iaa	7	4.8	0	0	0	7
aac(6′)-Iy	2	1.4	0	0	1	1
aph(3′)-Ia	5	3.4	0	5	0	0
aph(4)-Ia	11	7.5	0	11	0	0
bla$_{CTX-M-65}$	10	6.8	0	10	0	0
mcr-9.1	1	0.7	1	0	0	0
qacEDelta1	12	8.2	0	11	0	1
sul1	12	8.2	0	11	0	1
Total	71		1	59	1	10

[a] Antibiotic-resistant ontology name in accordance with the Comprehensive Antibiotic Resistance Database (CARD) software. [b] A total of 71 ARO counts were detected in 146 isolates.

Overall, 6.8% (10/146) MDR (resistance to 3 or more classes of antimicrobial agents, according to CARD classification) isolates were detected, of which 100% were recovered at broiler farms and belonged to serovar Infantis.

3.3. Virulence Profile

Overall, for the *Salmonella* strains from the four sources (hatcheries, farms, processing plants, and retail outlets), 4463 different virulence factors belonging to five virulence classes were identified. Genes classified as secretion systems and fimbrial adherence determinant classes accounted for the predominant virulence classes of 69.3% (3091/4463) and 29.2% (1302/4463) counts, respectively. Magnesium uptake, stress adaptation, and toxin classes accounted for less than 1.3% (56/4463) counts, respectively. *Salmonella* isolates ($n = 10$) recovered from the hatcheries contained virulence factors belonging to secretion systems (4.2%, 187/4463) and fimbrial adherence determinants (2.0%, 91/4463), whereas farm isolates ($n = 20$) were found to contain fimbrial adherence determinants, 4.4% (198/4463), and secretion system, 10.1% (451/4463). Processing plant *Salmonella* isolates ($n = 61$) contained predominantly factors in the secretion systems, fimbrial adherence determinants, and toxins, accounting for 30.0% (1341/4463), 12.4% (553/4463), and 0.6% (26/4463) count, respectively. Retail outlet isolates ($n = 55$) contained fimbrial adherence determinants, 10.3% (460/4463), secretion system, 24.9% (1112/4463), and toxin-related factors, 0.6% (27/4463). The differences in the detection of virulence factors among the sources were statistically significant ($p < 0.001$).

Serovars Kentucky, Javiana, and Infantis contained higher numbers of virulence factors (all related to secretion systems), accounting for 13.0% (578/4463), 12.1% (540/4463), and 12.1% (517/4463), respectively, of the virulence factors (Table S2). Therefore, it is pertinent to mention that they were the predominant serovars detected in this study.

S. Infantis isolates contained factors associated with secretion systems (TTSS-1 translocated effectors, TTSS-SPI-1-, and TTSS-SPI-2-encoded genes), 12.1% (540/4463), and factors associated with fimbrial adherence determinants (*bcfA, D, F, csg A, B, C, E, F, G, and lpfB, E*), 4.9% (220/4463). For the isolates of *S.* Javiana, 11.6% (517/4463), 4.4% (196/4463), and 0.6% (28/4463) were positive for factors associated with secretion systems, fimbrial adherence determinants (*bcfA, csgA, C, D, F, G,* and *fimF*), and toxins (*cdtB*), respectively. Only secretion system and fimbrial adherence determinant factors were detected among Kentucky isolates, accounting for 13.0% (578/4463) and 5.9% (263/4463), respectively. *S.* Schwarzengrund, Senftenberg, and Caracas contained predominantly factors associated with secretion systems at frequencies ranging from 1.3% to 2.8%. Seven serovars (Caracas, Chester, Enteritidis, Gaminara, Javiana, Montevideo, and Schwarzengrund) contained virulence factors related to toxins, where the *cdtB* was detected in all except serovar Enteritidis, where the *spvB* gene was detected.

3.4. Comparison of Frequency of Detection of Resistance and Virulence Factors in Salmonella Strains

Comparisons between the possession of virulence factors (VFDB accessions) and AMR genes (ARO accessions) across serovars were performed and detected 10 sources and years (farm-to-fork, hatcheries, processing plants, retail outlets, 'pluck shops', supermarkets, 2016, 2017, 2018, and 2019). Statistically significant positive correlations in *Salmonella* serovars isolated from farms, retail outlets, and 'pluck shops', as well as those isolated in 2016 and 2019 were detected, respectively ($p < 0.05$) (Figure 1). Negative and non-significant positive correlations are not displayed.

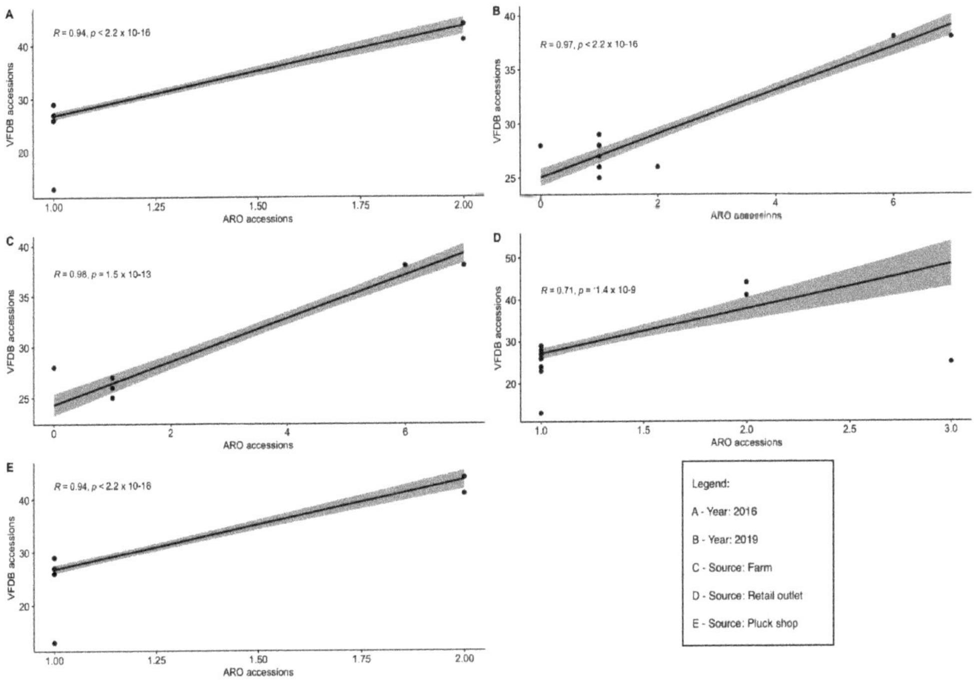

Figure 1. VFDB accessions (virulence) versus ARO accessions (AMR) (**A–E**).

3.5. Detection of ESβL Resistance Genes and Virulence Genes in Isolates of S. Infantis

A comparison of the phenotypic and genotypic resistance patterns in *S.* Infantis isolates is displayed in Table 4. The $bla_{CTX\text{-}M\text{-}65}$ gene was only detected among the *S.* Infantis isolates. Of the 10 isolates of serovar Infantis positive for $bla_{CTX\text{-}M\text{-}65}$ gene, phenotypically (using the disk diffusion method), two were resistant to two classes of antimicrobial agents, and six were MDR. However, genotypically, all 10 Infantis isolates exhibited MDR. Furthermore, the resistance gene *qacEDelta1* responsible for antiseptic resistance was found in all the 10 serovar Infantis isolates. Additionally, virulence factors associated with fimbrial adherence determinants and the secretion system were detected in all the 10 isolates of serovar Infantis.

Table 4. Detection of the $bla_{CTX\text{-}M\text{-}65}$ gene, other resistance genes, and virulence genes in *S.* Infantis.

BioSample	Isolate No. [a]	Phenotypic AMR Using the Disk Diffusion Method [b,c,d]							Genotypic Characteristics [e]	
		P	TE	CE	AM	PH	S	F	Other Resistance Genes Detected [f]	Virulence Factors
SAMN25867756	F 17	S	R	R	R	S	R	S	*qacEDelta1* *aph(4)-Ia* *aac(3)-IV* *sul1*	*agf/csg* *bcf* *lpf* TTSS (SPI-1 encode) TTSS (SPI-2 encode) TTSS-1 translocated effectors

Table 4. *Cont.*

BioSample	Isolate No. [a]	Phenotypic AMR Using the Disk Diffusion Method [b,c,d]							Genotypic Characteristics [e]	
		P	TE	CE	AM	PH	S	F	Other Resistance Genes Detected [f]	Virulence Factors
SAMN25867757	F 22	S	R	R	R	S	R	S	*qacEDelta1* *aph(4)-Ia* *aac(3)-IV* *sul1*	*agf/csg* *bcf* *lpf* TTSS (SPI-1 encode) TTSS (SPI-2 encode) TTSS-1 translocated effectors
SAMN14677229	F 11	S	R	R	R	S	R	S	*qacEDelta1* *aph(4)-Ia* *aac(3)-IV* *sul1*	*agf/csg* *bcf* *lpf* TTSS (SPI-1 encode) TTSS (SPI-2 encode) TTSS-1 translocated effectors
SAMN14677211	F 32	S	R	R	R	S	R	S	*aph(3′)-Ia* *qacEDelta1* *aph(4)-Ia* *aac(3)-IV* *sul1*	*agf/csg* *bcf* *lpf* TTSS (SPI-1 encode) TTSS (SPI-2 encode) TTSS-1 translocated effectors
SAMN14677232	F 36	S	R	R	R	S	R	S	*aph(3′)-Ia* *qacEDelta1* *aph(4)-Ia* *aac(3)-IV* *sul1*	*agf/csg* *bcf* *lpf* TTSS (SPI-1 encode) TTSS (SPI-2 encode) TTSS-1 translocated effectors
SAMN14677210	F 2	S	S	S	S	S	S	S	*qacEDelta1* *aph(4)-Ia* *aac(3)-IV* *sul1*	*agf/csg* *bcf* *lpf* TTSS (SPI-1 encode) TTSS (SPI-2 encode) TTSS-1 translocated effectors

Table 4. *Cont.*

BioSample	Isolate No.[a]	Phenotypic AMR Using the Disk Diffusion Method [b,c,d]							Genotypic Characteristics [e]	
		P	TE	CE	AM	PH	S	F	Other Resistance Genes Detected [f]	Virulence Factors
SAMN14677203	F 4	S	R	R	R	S	S	R	*qacEDelta1* *aph(4)-Ia* *aac(3)-IV* *sul1*	*agf/csg* *bcf* *lpf* TTSS (SPI-1 encode) TTSS (SPI-2 encode) TTSS-1 translocated effectors
SAMN14677209	UWI-F30	S	S	S	S	S	S	S	*aph(3′)-Ia* *qacEDelta1* *aph(4)-Ia* *aac(3)-IV* *sul1*	*agf/csg* *bcf* *lpf* TTSS (SPI-1 encode) TTSS (SPI-2 encode) TTSS-1 translocated effectors
SAMN14677207	UWI-F9	S	R	S	R	S	S	S	*qacEDelta1* *aph(4)-Ia* *aac(3)-IV* *sul1*	*agf/csg* *bcf* *lpf* TTSS (SPI-1 encode) TTSS (SPI-2 encode) TTSS-1 translocated effectors
SAMN14677208	UWI-F31	S	R	S	R	S	S	S	*aph(3′)-Ia* *qacEDelta1* *aph(4)-Ia* *aac(3)-IV* *sul1*	*agf/csg* *bcf* *lpf* TTSS (SPI-1 encode) TTSS (SPI-2 encode) TTSS-1 translocated effectors

[a] All 10 isolates were obtained from broiler farms comprising 7 (70%) cloacal swabs, 2 (20%) water supply (UWI-F30 and UWI-F9), and 1 drag swab of litter (UWI-F31) from where *blaCTX-M-65*, the only *ESβL*-resistance gene was detected. [b] P, penam (amoxicillin–clavulanic acid, 30 µg); TE, tetracycline (doxycycline, 30 µg); CE, cephalosporin (ceftriaxone, 30 µg); AM, aminoglycoside (gentamicin, 10 µg, and kanamycin, 30 µg); PH, phenicol (chloramphenicol, 30 µg); S, sulphonamide (sulfamethoxazole–trimethoprim, 23.75 and 1.25 µg); F, fluoroquinolone (ciprofloxacin, 5 µg). [c] A total of 146 (151 with controls) isolates of *Salmonella* were tested for AMR by the disk diffusion method, resistance genes, and virulence genes by WGS where 6.6% (10/151) were positive for *ESβL* resistance genes ($bla_{CTX\text{-}M\text{-}65}$). [d] S: Susceptible and R: Resistance. [e] Antimicrobial resistance and virulence analyses were performed using CARD and VFDB. [f] All isolates belonged to serovar Infantis and contained the *golS* gene, not shown.

4. Discussion

This is the first documented WGS study conducted in the poultry (broilers and layers) industry along the broiler production chain in Trinidad and Tobago, and the Caribbean region at large. Whole-genome sequencing analysis has been used to investigate genetic characteristics and phylogenies among *Salmonella* strains isolated from different origins, such as humans, food, animals, and the environment [33–36]. The current study was comprised of isolates from four cross-sectional studies conducted at the level of retail outlets (2016–2017) [25], broiler processing plants (2018) [23], broiler farms, and broiler hatcheries (2019) [22]. Although several limitations exist with the use of cross-sectional instead of longitudinal studies, this approach provides valuable information on the status of *Salmonella* shedding and contamination at the four levels (hatcheries, farms, processing plants, and retail outlets) of the broiler production chain in the country. Furthermore, the information obtained will lead to a better understanding of the epidemiology of *Salmonella* and the associated public health implications. Finally, this approach will also facilitate the implementation of an effective surveillance system across the poultry production system in the country.

Using the SISTR pipeline, a 76% agreement was detected with the traditional conventional serotyping method, which utilizes the White–Kauffman–Le Minor (WKL) scheme, which is based on immunological reactions to somatic (O) and flagellar (H) antigens [37]. However, it has been documented that conventional serotyping is time-consuming, labor-intensive, costly, and some isolates do not express serotype antigens due to a single nucleotide change in the genome [38,39]. On the other hand, the SISTR pipeline has been validated and a 94.6% overall serovar prediction accuracy was reported when 4291 genomes were analyzed [29]. In silico serotyping channels such as SISTR provide us with an understanding of the antigenic genes carried by an isolate and not necessarily what is expressed by that isolate, an advantage over traditional serotyping methods. In a study that compared three in silico pipelines, SISTR, SeqSero, and MLST to traditional serotyping techniques using a set of 813 verified clinical and laboratory isolates, 94.8%, 88.2%, and 88.3% accuracy, respectively was reported [40]. Of significance in our study was the incorrect serotype classification by the conventional method of potential public health important serotypes such as *S.* Albany, Senftenberg, Infantis, and Caracas. Variations in *Salmonella* serovars in poultry have been reported in different countries and are known to be a function of geographic location [5]. In Egypt [41], serovars Enteritidis and Typhimurium were isolated from broiler chickens at retail outlets; in Japan [42], serovars Infantis, Manhattan, Schwarzengrund from cecal samples in broilers; in China [43], serovars Pullorum, Gallinarum, Enteritidis, and Typhimurium were the predominant serotypes. Unlike our study where *S.* Kentucky, Javiana, Infantis, and Albany were the predominant serovars isolated along the broiler production chain.

Gene $bla_{\text{CTX-M-65}}$ was detected in 77% (31/40) of the *S.* Infantis strains isolated at Colombian processing plants [44], a finding comparable to the 50% (10/20) detected in the current study. Worldwide, the rapid development of resistance to extended-spectrum cephalosporins, predominantly associated with the production of β-lactamase-producing bacteria (EsβL) in different *Salmonella* serovars, has been reported. In agreement with our study, EsβL resistance genes have been detected in *Salmonella* strains isolated from animal products in several countries, including Korea (food animals and humans), 1.6% [45], Mexico (humans and animals), 6.6% [46], and Brazil (broiler chickens), where 27.8% [47] of isolates were positive for the $bla_{CTX\text{-}M}$ gene. The detection of 6.8% MDR strains among serovar Infantis isolates possessing the $bla_{CTX\text{-}M\text{-}65}$ gene is of public health significance due to the reported cross-transmission of EsβL-producing bacterial strains from poultry farms to other livestock farms and humans with the potential for wide-spread population infections [48,49]. A cause for concern is the detection of 10 MDR serovar Infantis strains, each possessing the $bla_{CTX\text{-}M\text{-}65}$ gene, in addition to 38 virulence homologs according to VFDB. In 2014, the detection of an MDR emergent *Salmonella* Infantis (ESI) strain, often containing the $bla_{CTX\text{-}M\text{-}65}$ gene, was first reported in Israel, and subsequently detected in Italy, Japan, and Russia [50–53]. However, retrospective sequencing tracked the origins of this clone to South America [54,55]. This ESI strain was documented to carry a large plasmid ESI (pESI) with several antimicrobial resistance, metal, and virulence genes. This clone was detected in retail meats in Tennessee, USA, in 2014, but by 2019 had spread throughout the USA to comprise 29% *Salmonella* isolated from retail chickens and 7% from retail turkey [56]. This clone also accounted for nearly 10% of all human Infantis cases by 2017 in the United States and was highly related to chicken sources [57]. The most frequently described ESβL genotype in Colombia between 1997–2018 was *CTX-M*, which was detected primarily in *S.* Typhimurium (40%; 65/164) and *S.* Infantis (29%; 48/164). Detection of bla_{CTX} genes has been reported in Latin American countries, such as Brazil and Argentina [58]. The assumption was that cephalosporin resistance development was due to the injection of ceftiofur into fertile eggs at hatcheries to prevent *E. coli*-induced omphalitis in day-old chicks [59]. This assumption was supported by a Canadian study that revealed a strong correlation between this practice and the increase in ceftiofur-resistant strains of *S.* Heidelberg [60]. This practice was not evident at hatcheries in our study, nor were *bla* genes detected among hatchery isolates.

A quick look into the NCBI Pathogen detection browser (https://www.ncbi.nlm.nih.gov/pathogens) allows us to determine that the eight strains of *S.* Infantis (ST32) detected in this study were highly related to the MDR emergent *S.* Infantis strains carrying $bla_{CTX\text{-}M\text{-}65}$ (https://www.ncbi.nlm.nih.gov/pathogens/tree#Salmonella/PDG000000002.2405/PDS000089910.160?term=CFSAN103822,%20CFSAN103806,%20CFSAN103805,%20CFSAN103797,%20CFSAN103801,%20CFSAN103796,%20CFSAN103798,%20CFSAN103802, accessed on 2 March 2022) reported in previous studies [50–57]. This highlights the usefulness of WGS approaches for AMR surveillance in a country or region, in this case, Trinidad and Tobago, considering the significant public health and clinical implications resulting from the presence of this large plasmid ESI. The plasmid detected in our *S.* Infantis carrying the CTX-M-65 gene (Accession: CP066336.1) contained 312,952 bp, differing from the plasmids reported in the USA [61] and Italy [52], which ranged from 316,160 to 323,122bp. These eight strains exhibited two resistance profiles: *aac(3)-IVa, aadA1, aph(3')-Ia, aph(4)-Ia,* $bla_{CTX\text{-}M\text{-}65}$*, dfrA14, gyrA_D87Y, mdsA, mdsB, sul1, tet(A)* (4 strains) and *aac(3)-IVa, aadA1, aph(4)-Ia,* $bla_{CTX\text{-}M\text{-}65}$*, dfrA14, gyrA_D87Y, mdsA, mdsB, sul1, tet(A)* (4 strains), according to the NCBI database (AMRFinderPlus). Our findings were similar to the *aph(4)-Ia, aac(3)-IVa, aph(3')-Ic,* $bla_{CTX\text{-}M\text{-}65}$*, fosA3, floR, dfrA14, gyrA_D87Y, sul1, tetA, aadA1* pattern detected in the USA [61] and *aph(4)-Ia, aac(3)-IVa, aph(3')-Ic,* $bla_{CTX\text{-}M\text{-}65}$*, fosA3, floR, dfrA14, sul1, tetA, aadA1* detected clinically in Italy [52], where both studies used ResFinder.

It must be highlighted that EsβL-producing *K. pneumoniae* was detected in 78.8% (41/52) of clinical isolates originating from a tertiary care hospital in Trinidad and Tobago, where the bla_{SHV} and $bla_{CTX\text{-}M}$ genes were predominantly detected [62]. It is of interest that all the MDR Infantis strains isolated in the current study originated from broiler farms. This is because there is a potential for *Salmonella* strains positive for bla_{CTX} gene, AMR, and associated virulence genes, to enter the human food chain through the processing plants and chicken products at the retail outlets. This is supported by reports documenting close association of MDR Infantis strains recovered from the broiler population to animal production environments, eventually spreading into the food chain and potentially humans [63,64].

As with this study, aminoglycoside resistance genes and *sul1* genes were also detected in *Salmonella* Infantis strains isolated in a recent study conducted at three Colombian broiler processing plants [44]. Sulphonamide resistance conferred by *sul* genes [65] was reported in Canadian swine and chicken *Salmonella* isolates [66] and at a broiler processing plant in China [67]. However, in our study, only the *sul1* gene was detected in all our *S.* Infantis strains and the only Senftenberg strain assessed. Arkali et al. [68] detected the *sul1* gene among 58% of *Salmonella* isolated from chickens in Eastern Turkey. The detection of one mobilized colistin resistance *mcr-9.1* gene [69] in an isolate of serovar Senftenberg was not a significant finding. This gene is not associated with colistin resistance in *Salmonella* or *E. coli* in the United States [70]. However, detecting this novel *mcr-9* homolog is crucial as it can confer phenotypic resistance to colistin and warrants close monitoring [16].

The *qacEdelta1* gene, known to confer resistance to antiseptics, was also detected in *Salmonella* from retail foods of animal origin [20]. It must be considered that the presence of antimicrobial resistance genes can represent the phenotypic resistance of antimicrobial agents, and thus diminish their effectiveness when used on farms or processing plants. However, it is important to mention that there are several mechanisms of antimicrobial resistance in bacteria. It is not always associated with a specific gene responsible for resistance. This supports our findings where resistance genes were found in two Infantis strains, but they were all sensitive phenotypically. Cross-resistance to antimicrobial agents can occur with resistance within group members of chemical-related compounds, and/or with a similar mechanism of action [71,72]. The correlation of genotypic and phenotypic resistance was variable in our study, contrary to the findings of other studies where the harmonic correlation was evident [67,73]. The lack of correlation between phenotypic and genotypic resistance profiles may occur due to the low sensitivity and specificity of the disk method, inoculum concentration, laboratory capacity, and individual skill. Misalignments between phenotypic and genotypic resistance patterns have been reported by others [74,75].

In the current study, only 6.8% (10/146) of the isolates, based on genotypic characterization, exhibited multidrug resistance, at variance with the 96.6% reported in *Salmonella* isolated from chickens sampled at chicken farms in South Africa [76] and the 27.3% reported for *Salmonella* strains isolated from broilers in Egypt [77]. Therefore, our low frequency of detecting MDR is of therapeutic significance at the broiler farm level in the country.

SPI-1 and *SPI-2* genes enable the invasion of eukaryotic cells, induction of macrophage cytotoxicity, invasion of phagocytes, and survival inside phagocytic cells [78–81]. The inactivation of the TTSS-1 translocated effector gene *sipB in S.* Dublin has been associated with reduced fluid secretion and inflammation [82]. This is of public health significance because of the 73 genes detected in the current study, 49.3% and 35.6% were detected in *Salmonella* strains isolated from processing plants and retail outlets, respectively, highlighting the risk posed to consumers should they be infected with a serovar positive for the gene. In the current study, serovars Aberdeen, Anatum, Enteritidis, Infantis, Javiana, Manhattan, Virchow, and Weltevreden were all positive for the *sipB* gene.

Fimbrial adherence factors that aid intestinal adhesion such as long polar fimbriae (*lpfA*) and aggregative fimbriae (*agfA/csgA*) are highly conserved in *Salmonella* and have been implicated in biofilm formation and adhesion to surfaces and epithelial cells that is an important stage before biofilm formation, respectively [83,84]. This is important in the current study because 99.3% of the isolates were positive for the *csgA* gene, therefore having the potential for biofilm formation and persistence in the environment. In addition, the high incidence of *csgA* in our study is comparable to the findings in different serovars as reported by others [85,86].

Typhoid toxin/*cdtB* cytolethal distending toxin B, previously thought to be a unique virulence factor in *S.* Typhi, was recently characterized in at least 40 non-typhoidal *Salmonella* serovars [87], as evident in our study. The detection of virulence genes *invA*, *csgA*, *lpfA*, *sopE*, and *spvC* in our *S.* Enteritidis strains agrees with the findings of studies conducted on chickens sold at Bangladeshi retail outlets [88], in food and humans in Brazil [89], and in humans and animals in Iran [90].

The positive correlations in the detection of AMR and virulence genes in the *Salmonella* serovars isolated from farms, retail outlets, and 'pluck shops' are indicative of close similarities in the occurrence of AMR and virulence genes in different serovars and isolates in the study area or source-dependent AMR/virulence profiles. The presence of virulence genes and the occurrence of AMR *Salmonella* isolates can potentially accelerate the pathogenicity of microbes [91]. It has also been reported that the emergence of resistant *Salmonella enterica* solely depends on genetic and pathogenicity mechanisms that may enhance survivability by preserving their drug resistance genes [92]. However, the correlation between AMR and virulence has been shown to vary in studies conducted by others. The acquisition of AMR by *Salmonella* isolates decreases [93,94], increases [95,96], or does not change [97,98] their potential virulence according to those authors.

5. Conclusions

This study highlighted the antimicrobial resistance and virulence genes associated with *Salmonella* serovars isolated along the broiler production chain in Trinidad and Tobago. The detection of the $bla_{CTX\text{-}M\text{-}65}$ gene, MDR, and highly virulent *S.* Infantis isolates based on their genotypes, is cause for concern given their international emergence and implications for human health. The positive correlation of resistance and virulence genes detected at broiler farms, processing plants, and retail outlets ('pluck shops') is significant since the latter two stages of the broiler continuum can directly impact consumers of contaminated, improperly handled, or cooked chicken.

The availability of these genomes will help future source tracking during epidemiological investigations associated with *Salmonella* foodborne outbreaks in the region and worldwide. Therefore, the abundance of data from several sources in the country will benefit the scientific community at large.

Supplementary Materials: The following are available online at https://www.mdpi.com/article/10.3390/microorganisms10030570/s1, Table S1: Metadata of 146 *Salmonella* isolates detected along the broiler production chain in Trinidad and Tobago. Table S2: Distribution of virulence genes and classes among the various serovars of *Salmonella* isolated in this study.

Author Contributions: Conceptualization, A.A.A., A.S.K. and K.G.; methodology, A.A.A., A.S.K., R.E.P., N.G.-E., M.M., C.L., G.H.T. and S.A.; software, R.E.P., N.G.-E., M.M. and G.H.T.; validation, R.E.P., N.G.-E., M.M. and G.H.T.; formal analysis, R.E.P., A.A.A. and A.S.K.; investigation, A.A.A., A.S.K. and K.G.; resources, A.A.A., N.G.-E., M.M., G.H.T. and S.A.; data curation, A.A.A., A.S.K. and R.E.P.; writing—original draft preparation, A.A.A., A.S.K. and W.A.; writing—review and editing, A.A.A., A.S.K., K.G., R.E.P., N.G.-E., M.M., C.L., G.H.T. and S.A.; visualization, R.E.P., A.S.K. and A.A.A.; supervision, A.A.A., K.G. and W.A.; project administration, A.A.A., A.S.K. and K.G.; funding acquisition, A.A.A. and A.S.K. All authors have read and agreed to the published version of the manuscript.

Funding: The University of the West Indies, St. Augustine Campus Research and Publication Fund Committee approved the funding for the project (Research grant #2660-457522 on 3 July 2017).

Institutional Review Board Statement: The study was approved and conducted under terms approved by the University of the West Indies, St. Augustine Campus Research Committee (Research grant #2660-457522 on 13 April 2016). The UWI St. Augustine Campus Ethics Committee exempted the project from ethical review after assessing the research proposal.

Informed Consent Statement: Written informed consent has been obtained from the patient(s) by the Caribbean Public Health Agency (CARPHA) from where the five isolates of *Salmonella* originated, to publish this paper.

Data Availability Statement: All the data are contained within the article and the Supplementary Materials.

Acknowledgments: We thank the staff of the Public Health Agency of Canada (PHAC) Laboratory for conducting some whole-genome sequencing. We are grateful to the staff of the Bacteriology section of the Caribbean Public Health Agency (CARPHA), Port-of-Spain for providing strains of *Salmonella* Infantis and Enteritidis isolated from human gastroenteritis cases in the country. The technical assistance of Alva Marie Stewart-Johnson and Sannandan Samlal is appreciated.

Conflicts of Interest: The authors declare no conflict of interest. The funders had no role in the design of the study; in the collection, analysis, or interpretation of data; in the writing of the manuscript, or in the decision to publish the results. The views expressed in this article are those of the authors and do not necessarily reflect the official policy of the Department of Health and Human Services, the U.S. Food and Drug Administration, or the U.S. Government.

References

1. Ulaya, D.W. Determination of virulence factors in *Salmonella* isolates of human, poultry, and dog origin in Lusaka district, Zambia. Master's Thesis, University of Zambia, Lusaka, Zambia, 2014.

2. Centers for Disease Control and Prevention. *Salmonella*. Available online: https://www.cdc.gov/salmonella/index.html (accessed on 3 October 2020).
3. European Food Safety Authority; European Centre for Disease Prevention Control. The European Union One Health 2018 zoonoses report. *EFSA J.* **2019**, *17*, e05926. [CrossRef]
4. Shilangale, R.P.; Di Giannatale, E.; Chimwamurombe, P.M.; Kaaya, G.P. Prevalence and antimicrobial resistance pattern of *Salmonella* in animal feed produced in Namibia. *Vet. Ital.* **2012**, *48*, 125–132. [PubMed]
5. Foley, S.L.; Lynne, A.M.; Nayak, R. *Salmonella* challenges: Prevalence in swine and poultry and potential pathogenicity of such isolates. *J. Anim. Sci.* **2008**, *86*, E149–E162. [CrossRef] [PubMed]
6. Kimera, Z.I.; Mshana, S.E.; Rweyemamu, M.M.; Mboera, L.E.G.; Matee, M.I.N. Antimicrobial use and resistance in food-producing animals and the environment: An African perspective. *Antimicrob. Resist. Infect. Control.* **2020**, *9*, 37. [CrossRef] [PubMed]
7. Ma, F.; Xu, S.; Tang, Z.; Li, Z.; Zhang, L. Use of antimicrobials in food animals and impact of transmission of antimicrobial resistance on humans. *Biosaf. Health* **2021**, *3*, 32–38. [CrossRef]
8. World Health Organization (WHO). *Critically Important Antimicrobials for Human Medicine*; World Health Organisation: Geneva, Switzerland, 2019.
9. Collignon, P.J.; Conly, J.M.; Andremont, A.; McEwen, S.A.; Aidara-Kane, A.; Agerso, Y.; Andremont, A.; Collignon, P.; Conly, J.; Dang Ninh, T.; et al. World Health Organization ranking of antimicrobials according to their importance in human medicine: A critical step for developing risk management strategies to control antimicrobial resistance from food animal production. *Clin. Infect. Dis.* **2016**, *63*, 1087–1093. [CrossRef] [PubMed]
10. Wang, M.; Sahm, D.F.; Jacoby, G.A.; Hooper, D.C. Emerging plasmid-mediated quinolone resistance associated with the qnr gene in *Klebsiella pneumoniae* clinical isolates in the United States. *Antimicrob. Agents Chemother.* **2004**, *48*, 1295–1299. [CrossRef]
11. Hooper, D.C.; Jacoby, G.A. Mechanisms of drug resistance: Quinolone resistance. *Ann. NY Acad. Sci.* **2015**, *1354*, 12–31. [CrossRef]
12. Fàbrega, A.; Madurga, S.; Giralt, E.; Vila, J. Mechanism of action of and resistance to quinolones. *Microb. Biotechnol.* **2009**, *2*, 40–61. [CrossRef]
13. Chen, S.; Zhao, S.; White, D.G.; Schroeder, C.M.; Lu, R.; Yang, H.; McDermott, P.F.; Ayers, S.; Meng, J. Characterization of multiple-antimicrobial-resistant *Salmonella* serovars isolated from retail meats. *Appl. Environ. Microbiol.* **2004**, *70*, 1–7. [CrossRef]
14. Miko, A.; Pries, K.; Schroeter, A.; Helmuth, R. Molecular mechanisms of resistance in multidrug-resistant serovars of *Salmonella enterica* isolated from foods in Germany. *J. Antimicrob. Chemother.* **2005**, *56*, 1025–1033. [CrossRef] [PubMed]
15. Liu, Y.Y.; Wang, Y.; Walsh, T.R.; Yi, L.X.; Zhang, R.; Spencer, J.; Doi, Y.; Tian, G.; Dong, B.; Huang, X.; et al. Emergence of plasmid-mediated colistin resistance mechanism *mcr-1* in animals and human beings in China: A microbiological and molecular biological study. *Lancet Infect. Dis.* **2016**, *16*, 161–168. [CrossRef]
16. Carroll, L.M.; Gaballa, A.; Guldimann, C.; Sullivan, G.; Henderson, L.O.; Wiedmann, M. Identification of novel mobilized colistin resistance gene *mcr-9* in a multidrug-resistant, colistin-susceptible *Salmonella enterica* serotype Typhimurium isolate. *mBio* **2019**, *10*, e00853-19. [CrossRef] [PubMed]
17. Borowiak, M.; Baumann, B.; Fischer, J.; Thomas, K.; Deneke, C.; Hammerl, J.A.; Szabo, I.; Malorny, B. Development of a novel mcr-6 to mcr-9 multiplex PCR and assessment of mcr-1 to mcr-9 occurrence in colistin-resistant *Salmonella enterica* isolates from environment, feed, animals and food (2011–2018) in Germany. *Front. Microbiol.* **2020**, *11*, 80. [CrossRef]
18. Rebelo, A.R.; Bortolaia, V.; Kjeldgaard, J.S.; Pedersen, S.K.; Leekitcharoenphon, P.; Hansen, I.M.; Guerra, B.; Malorny, B.; Borowiak, M.; Hammerl, J.A.; et al. Multiplex PCR for detection of plasmid-mediated colistin resistance determinants, *mcr-1*, *mcr-2*, *mcr-3*, *mcr-4*, and *mcr-5* for surveillance purposes. *Eurosurveillance* **2018**, *23*, 17–00672. [CrossRef]
19. Eguale, T.; Birungi, J.; Asrat, D.; Njahira, M.N.; Njuguna, J.; Gebreyes, W.A.; Gunn, J.S.; Djikeng, A.; Engidawork, E. Genetic markers associated with resistance to beta-lactam and quinolone antimicrobials in non-typhoidal *Salmonella* isolates from humans and animals in central Ethiopia. *Antimicrob. Resist. Infect. Control.* **2017**, *6*, 13. [CrossRef]
20. Deng, W.; Quan, Y.; Yang, S.; Guo, L.; Zhang, X.; Liu, S.; Chen, S.; Zhou, K.; He, L.; Li, B.; et al. Antibiotic resistance in *Salmonella* from retail foods of animal origin and its association with disinfectant and heavy metal resistance. *Microb. Drug Resist.* **2018**, *24*, 782–791. [CrossRef]
21. Vilela, F.P.; Gomes, C.N.; Passaglia, J.; Rodrigues, D.P.; Costa, R.G.; Tiba Casas, M.R.; Fernandes, S.A.; Falcão, J.P.; Campioni, F. Genotypic resistance to quinolone and tetracycline in *Salmonella* Dublin strains isolated from humans and animals in Brazil. *Microb. Drug Resist.* **2019**, *25*, 143–151. [CrossRef]
22. Khan, A.S.; Georges, K.; Rahaman, S.; Abebe, W.; Adesiyun, A.A. Occurrence, risk factors, serotypes and antimicrobial resistance of *Salmonella* strains isolated from imported fertile hatching eggs, hatcheries, and broiler farms in Trinidad and Tobago. *J. Food Prot.* **2021**, *85*, 266–277. [CrossRef]
23. Khan, A.S.; Georges, K.; Rahaman, S.; Abebe, W.; Adesiyun, A.A. Characterization of *Salmonella* isolates recovered from stages of the processing lines at four broiler processing plants in Trinidad and Tobago. *Microorganisms* **2021**, *9*, 1048. [CrossRef]
24. Khan, A.S.; Georges, K.; Rahaman, S.; Abdela, W.; Adesiyun, A.A. Antimicrobial resistance of *Salmonella* isolates recovered from chickens sold at retail outlets in Trinidad. *J. Food Prot.* **2018**, *81*, 1880–1889. [CrossRef] [PubMed]
25. Khan, A.S.; Georges, K.; Rahaman, S.; Abdela, W.; Adesiyun, A.A. Prevalence and serotypes of *Salmonella* spp. on chickens sold at retail outlets in Trinidad. *PLoS ONE* **2018**, *13*, e0202108. [CrossRef] [PubMed]
26. Rodrigo, S.; Adesiyun, A.A.; Asgarali, Z.; Swanston, W. Occurrence of selected foodborne pathogens on poultry and poultry giblets from small retail processing operations in Trinidad. *J. Food Prot.* **2006**, *69*, 1096–1105. [CrossRef] [PubMed]

27. Andrews, W. Manual of food quality control. 4. Rev. 1. Microbiological analysis. Food and Drug Administration. *FAO Food Nutr. Pap.* **1992**, *14*, 1–338.
28. Bankevich, A.; Nurk, S.; Antipov, D.; Gurevich, A.A.; Dvorkin, M.; Kulikov, A.S.; Lesin, V.M.; Nikolenko, S.I.; Pham, S.; Prjibelski, A.D.; et al. SPAdes: A new genome assembly algorithm and its applications to single-cell sequencing. *J. Comput. Biol.* **2012**, *19*, 455–477. [CrossRef]
29. Yoshida, C.E.; Kruczkiewicz, P.; Laing, C.R.; Lingohr, E.J.; Gannon, V.P.; Nash, J.H.; Taboada, E.N. The *Salmonella* in silico typing resource (SISTR): An open web-accessible tool for rapidly typing and subtyping draft *Salmonella* genome assemblies. *PLoS ONE* **2016**, *11*, e0147101. [CrossRef]
30. Hyatt, D.; Chen, G.L.; Locascio, P.F.; Land, M.L.; Larimer, F.W.; Hauser, L.J. Prodigal: Prokaryotic gene recognition and translation initiation site identification. *BMC Bioinform.* **2010**, *11*, 119. [CrossRef]
31. Liu, B.; Zheng, D.; Jin, Q.; Chen, L.; Yang, J. VFDB 2019: A comparative pathogenomic platform with an interactive web interface. *Nucleic Acids Res.* **2019**, *47*, D687–D692. [CrossRef]
32. Alcock, B.P.; Raphenya, A.R.; Lau, T.T.Y.; Tsang, K.K.; Bouchard, M.; Edalatmand, A.; Huynh, W.; Nguyen, A.V.; Cheng, A.A.; Liu, S.; et al. CARD 2020: Antibiotic resistome surveillance with the comprehensive antibiotic resistance database. *Nucleic. Acids Res.* **2020**, *48*, D517–D525. [CrossRef]
33. Lee, D.H.; Hyeon, J.Y.; Kim, J.; Kim, J.S.; Kim, S.J.; Jeon, S.E.; Choi, S.W.; Hong, W.T.; Song, C.S.; Lee, S.W. Close genetic relationship between *Salmonella enterica* serovar Enteritidis isolated from patients with diarrhoea and poultry in the Republic of Korea. *Clin. Microbiol. Infect.* **2015**, *21*, e68–e70. [CrossRef]
34. Almeida, F.; Medeiros, M.I.; Kich, J.D.; Falcao, J.P. Virulence-associated genes, antimicrobial resistance and molecular typing of *Salmonella* Typhimurium strains isolated from swine from 2000 to 2012 in Brazil. *J. Appl. Microbiol.* **2016**, *120*, 1677–1690. [CrossRef] [PubMed]
35. Pearce, M.E.; Alikhan, N.F.; Dallman, T.J.; Zhou, Z.; Grant, K.; Maiden, M.C.J. Comparative analysis of core genome MLST and SNP typing within a European *Salmonella* serovar Enteritidis outbreak. *Int. J. Food. Microbiol.* **2018**, *274*, 1–11. [CrossRef] [PubMed]
36. Pightling, A.W.; Pettengill, J.B.; Luo, Y.; Baugher, J.D.; Rand, H.; Strain, E. Interpreting whole-genome sequence analyses of foodborne bacteria for regulatory applications and outbreak investigations. *Front. Microbiol.* **2018**, *9*, 1482. [CrossRef] [PubMed]
37. Guibourdenche, M.; Roggentin, P.; Mikoleit, M.; Fields, P.I.; Bockemuhl, J.; Grimont, P.A.; Weill, F.X. Supplement 2003–2007 (No. 47) to the White-Kauffmann-Le Minor scheme. *Res. Microbiol.* **2010**, *161*, 26–29. [CrossRef]
38. Li, P.; Liu, Q.; Luo, H.; Liang, K.; Yi, J.; Luo, Y.; Hu, Y.; Han, Y.; Kong, Q. O-serotype conversion in *Salmonella* Typhimurium induces protective immune responses against invasive non-typhoidal *Salmonella* infections. *Front. Immunol.* **2017**, *8*, 1647. [CrossRef]
39. Chui, L.; Ferrato, C.; Li, V.; Christianson, S. Comparison of molecular and in silico *Salmonella* serotyping for *Salmonella* surveillance. *Microorganisms* **2021**, *9*, 955. [CrossRef]
40. Yachison, C.A.; Yoshida, C.; Robertson, J.; Nash, J.H.E.; Kruczkiewicz, P.; Taboada, E.N.; Walker, M.; Reimer, A.; Christianson, S.; Nichani, A.; et al. The validation and implications of using whole-genome sequencing as a replacement for traditional serotyping for a National *Salmonella* reference laboratory. *Front. Microbiol.* **2017**, *8*, 1044. [CrossRef]
41. Elkenany, R.; Elsayed, M.M.; Zakaria, A.I.; El-sayed, S.A.-E.-S.; Rizk, M.A. Antimicrobial resistance profiles and virulence genotyping of *Salmonella enterica* serovars recovered from broiler chickens and chicken carcasses in Egypt. *BMC Vet. Res.* **2019**, *15*, 124. [CrossRef]
42. Duc, V.M.; Nakamoto, Y.; Fujiwara, A.; Toyofuku, H.; Obi, T.; Chuma, T. Prevalence of *Salmonella* in broiler chickens in Kagoshima, Japan in 2009 to 2012 and the relationship between serovars changing and antimicrobial resistance. *BMC Vet. Res.* **2019**, *15*, 108. [CrossRef]
43. Wang, X.; Wang, H.; Li, T.; Liu, F.; Cheng, Y.; Guo, X.; Wen, G.; Luo, Q.; Shao, H.; Pan, Z.; et al. Characterization of *Salmonella* spp. isolated from chickens in Central China. *BMC Vet. Res.* **2020**, *16*, 299. [CrossRef]
44. Ramirez-Hernandez, A.; Carrascal-Camacho, A.K.; Varon-Garcia, A.; Brashears, M.M.; Sanchez-Plata, M.X. Genotypic characterization of antimicrobial resistant *Salmonella* spp. strains from three poultry processing plants in Colombia. *Foods* **2021**, *10*, 491. [CrossRef] [PubMed]
45. Tamang, M.D.; Nam, H.M.; Kim, T.S.; Jang, G.C.; Jung, S.C.; Lim, S.K. Emergence of extended-spectrum beta-lactamase (*CTX-M-15* and *CTX-M-14*)-producing nontyphoid *Salmonella* with reduced susceptibility to ciprofloxacin among food animals and humans in Korea. *J. Clin. Microbiol.* **2011**, *49*, 2671–2675. [CrossRef] [PubMed]
46. Aguilar-Montes de Oca, S.; Talavera-Rojas, M.; Soriano-Vargas, E.; Barba-Leon, J.; Vazquez-Navarrete, J.; Acosta-Dibarrat, J.; Salgado-Miranda, C. Phenotypic and genotypic profile of clinical and animal multidrug-resistant *Salmonella enterica* isolates from Mexico. *J. Appl. Microbiol.* **2018**, *124*, 67–74. [CrossRef] [PubMed]
47. Mendonça, E.P.; Melo, R.T.; Oliveira, M.R.M.; Monteiro, G.P.; Peres, P.A.B.M.; Fonseca, B.B.; Giombelli, A.; Rossi, D.A. Characteristics of virulence, resistance and genetic diversity of strains of *Salmonella* Infantis isolated from broiler chicken in Brazil. *Braz. Vet. Res.* **2020**, *40*, 29–38. [CrossRef]
48. Brinas, L.; Moreno, M.A.; Zarazaga, M.; Porrero, C.; Saenz, Y.; Garcia, M.; Dominguez, L.; Torres, C. Detection of *CMY-2*, *CTX-M-14*, and *SHV-12* beta-lactamases in *Escherichia coli* fecal-sample isolates from healthy chickens. *Antimicrob. Agents Chemother.* **2003**, *47*, 2056–2058. [CrossRef]

49. Winokur, P.L.; Brueggemann, A.; DeSalvo, D.L.; Hoffmann, L.; Apley, M.D.; Uhlenhopp, E.K.; Pfaller, M.A.; Doern, G.V. Animal and human multidrug-resistant, cephalosporin-resistant *Salmonella* isolates expressing a plasmid-mediated *CMY-2 AmpC* beta-lactamase. *Antimicrob. Agents Chemother.* **2000**, *44*, 2777–2783. [CrossRef]
50. Aviv, G.; Tsyba, K.; Steck, N.; Salmon-Divon, M.; Cornelius, A.; Rahav, G.; Grassl, G.A.; Gal-Mor, O. A unique megaplasmid contributes to stress tolerance and pathogenicity of an emergent *Salmonella enterica* serovar Infantis strain. *Environ. Microbiol.* **2014**, *16*, 977–994. [CrossRef]
51. Bogomazova, A.N.; Gordeeva, V.D.; Krylova, E.V.; Soltynskaya, I.V.; Davydova, E.E.; Ivanova, O.E.; Komarov, A.A. Mega-plasmid found worldwide confers multiple antimicrobial resistance in *Salmonella* Infantis of broiler origin in Russia. *Int. J. Food Microbiol.* **2020**, *319*, 108497. [CrossRef]
52. Franco, A.; Leekitcharoenphon, P.; Feltrin, F.; Alba, P.; Cordaro, G.; Iurescia, M.; Tolli, R.; D'Incau, M.; Staffolani, M.; Di Giannatale, E.; et al. Emergence of a clonal lineage of multidrug-resistant *ESβL*-producing *Salmonella* Infantis transmitted from broilers and broiler meat to humans in Italy between 2011 and 2014. *PLoS ONE* **2015**, *10*, e0144802. [CrossRef]
53. Yokoyama, E.; Ando, N.; Ohta, T.; Kanada, A.; Shiwa, Y.; Ishige, T.; Murakami, K.; Kikuchi, T.; Murakami, S. A novel subpopulation of *Salmonella enterica* serovar Infantis strains isolated from broiler chicken organs other than the gastrointestinal tract. *Vet. Microbiol.* **2015**, *175*, 312–318. [CrossRef]
54. Cartelle Gestal, M.; Zurita, J.; Paz, Y.M.A.; Ortega-Paredes, D.; Alcocer, I. Characterization of a small outbreak of *Salmonella enterica* serovar Infantis that harbour *CTX-M-65* in Ecuador. *Braz. J. Infect. Dis.* **2016**, *20*, 406–407. [CrossRef] [PubMed]
55. Palma, N.; Pons, M.J.; Gomes, C.; Mateu, J.; Riveros, M.; Garcia, W.; Jacobs, J.; Garcia, C.; Ochoa, T.J.; Ruiz, J. Resistance to quinolones, cephalosporins and macrolides in *Escherichia coli* causing bacteraemia in Peruvian children. *J. Glob. Antimicrob. Resist.* **2017**, *11*, 28–33. [CrossRef] [PubMed]
56. Tyson, G.H.; Li, C.; Harrison, L.B.; Martin, G.; Hsu, C.H.; Tate, H.; Tran, T.T.; Strain, E.; Zhao, S. A multidrug-resistant *Salmonella* Infantis clone is spreading and recombining in the United States. *Microb. Drug Resist.* **2021**, *27*, 792–799. [CrossRef] [PubMed]
57. Brown, A.C.; Chen, J.C.; Watkins, L.K.F.; Campbell, D.; Folster, J.P.; Tate, H.; Wasilenko, J.; Van Tubbergen, C.; Friedman, C.R. *CTX-M-65* extended-spectrum beta-lactamase-producing *Salmonella enterica* serotype Infantis, United States. *Emerg. Infect. Dis.* **2018**, *24*, 2284–2291. [CrossRef] [PubMed]
58. Bonnet, R. Growing group of extended-spectrum beta-lactamases: The *CTX-M* enzymes. *Antimicrob. Agents Chemother.* **2004**, *48*, 1–14. [CrossRef]
59. Hofacre, C.L.; Fricke, J.A.; Inglis, T. Antimicrobial drug use in poultry. In *Antimicrobial Therapy in Veterinary Medicine*; Giguère, S., Prescott, J.F., Dowling, P.M., Eds.; John Wiley & Sons: Hoboken, NJ, USA, 2013; pp. 569–587.
60. Dutil, L.; Irwin, R.; Finley, R.; Ng, L.K.; Avery, B.; Boerlin, P.; Bourgault, A.M.; Cole, L.; Daignault, D.; Desruisseau, A.; et al. Ceftiofur resistance in *Salmonella enterica* serovar Heidelberg from chicken meat and humans, Canada. *Emerg. Infect. Dis.* **2010**, *16*, 48–54. [CrossRef]
61. Tate, H.; Folster, J.P.; Hsu, C.H.; Chen, J.; Hoffmann, M.; Li, C.; Morales, C.; Tyson, G.H.; Mukherjee, S.; Brown, A.C.; et al. Comparative analysis of extended-spectrum-beta-lactamase CTX-M-65-producing *Salmonella enterica* serovar Infantis isolates from humans, food animals, and retail chickens in the United States. *Antimicrob. Agents Chemother.* **2017**, *61*, e00488-17. [CrossRef]
62. Cheddie, P.; Dziva, F.; Akpaka, P.E. Detection of a CTX-M group 2 beta-lactamase gene in a *Klebsiella pneumoniae* isolate from a tertiary care hospital, Trinidad and Tobago. *Ann. Clin. Microbiol. Antimicrob.* **2017**, *16*, 33. [CrossRef]
63. Hindermann, D.; Gopinath, G.; Chase, H.; Negrete, F.; Althaus, D.; Zurfluh, K.; Tall, B.D.; Stephan, R.; Nuesch-Inderbinen, M. *Salmonella enterica* serovar Infantis from food and human infections, Switzerland, 2010–2015: Poultry-related multidrug resistant clones and an emerging ESBL producing clonal lineage. *Front. Microbiol.* **2017**, *8*, 1322. [CrossRef]
64. Nógrády, N.; Király, M.; Davies, R.; Nagy, B. Multidrug resistant clones of *Salmonella* Infantis of broiler origin in Europe. *Int. J. Food Microbiol.* **2012**, *157*, 108–112. [CrossRef]
65. Sköld, O. Resistance to trimethoprim and sulfonamides. *Vet. Res.* **2001**, *32*, 261–273. [CrossRef] [PubMed]
66. Kozak, G.K.; Pearl, D.L.; Parkman, J.; Reid-Smith, R.J.; Deckert, A.; Boerlin, P. Distribution of sulfonamide resistance genes in *Escherichia coli* and *Salmonella* isolates from swine and chickens at abattoirs in Ontario and Quebec, Canada. *Appl. Environ. Microbiol.* **2009**, *75*, 5999–6001. [CrossRef] [PubMed]
67. Zhu, Y.; Lai, H.; Zou, L.; Yin, S.; Wang, C.; Han, X.; Xia, X.; Hu, K.; He, L.; Zhou, K.; et al. Antimicrobial resistance and resistance genes in *Salmonella* strains isolated from broiler chickens along the slaughtering process in China. *Int. J. Food Microbiol.* **2017**, *259*, 43–51. [CrossRef] [PubMed]
68. Arkali, A.; Cetinkaya, B. Molecular identification and antibiotic resistance profiling of *Salmonella* species isolated from chickens in eastern Turkey. *BMC Vet. Res.* **2020**, *16*, 205. [CrossRef] [PubMed]
69. Maguire, M.; Khan, A.S.; Adesiyun, A.A.; Georges, K.; Gonzalez-Escalona, N. Closed genome sequence of a *Salmonella enterica* serotype Senftenberg strain carrying the *mcr-9* gene isolated from broken chicken eggshells in Trinidad and Tobago. *Microbiol. Resour. Announc.* **2021**, *10*, e0146520. [CrossRef]
70. Tyson, G.H.; Li, C.; Hsu, C.H.; Ayers, S.; Borenstein, S.; Mukherjee, S.; Tran, T.T.; McDermott, P.F.; Zhao, S. The *mcr-9* Gene of *Salmonella* and *Escherichia coli* is not associated with colistin resistance in the United States. *Antimicrob. Agents Chemother.* **2020**, *64*, e00573-20. [CrossRef]
71. Sefton, A.M. Mechanisms of antimicrobial resistance: Their clinical relevance in the new millennium. *Drugs* **2002**, *62*, 557–566. [CrossRef]

72. Sosa, A.d.J.; Byarugaba, D.K.; Amábile-Cuevas, C.F.; Hsueh, P.-R.; Kariuki, S.; Okeke, I.N. *Antimicrobial Resistance in Developing Countries*; Springer: Berlin, Germany, 2010; pp. 1–554.
73. Alam, S.B.; Mahmud, M.; Akter, R.; Hasan, M.; Sobur, A.; Nazir, K.N.H.; Noreddin, A.; Rahman, T.; El Zowalaty, M.E.; Rahman, M. Molecular detection of multidrug resistant *Salmonella* species isolated from broiler farm in Bangladesh. *Pathogens* **2020**, *9*, 201. [CrossRef]
74. Van, T.T.; Chin, J.; Chapman, T.; Tran, L.T.; Coloe, P.J. Safety of raw meat and shellfish in Vietnam: An analysis of *Escherichia coli* isolations for antibiotic resistance and virulence genes. *Int. J. Food Microbiol.* **2008**, *124*, 217–223. [CrossRef]
75. Paterson, D.L.; Bonomo, R.A. Extended-spectrum beta-lactamases: A clinical update. *Clin. Microbiol. Rev.* **2005**, *18*, 657–686. [CrossRef]
76. Akinola, S.A.; Mwanza, M.; Ateba, C.N. Occurrence, genetic diversities and antibiotic resistance profiles of *Salmonella* serovars isolated from chickens. *Infect. Drug Resist.* **2019**, *12*, 3327–3342. [CrossRef] [PubMed]
77. Ammar, A.M.; Abdeen, E.E.; Abo-Shama, U.H.; Fekry, E.; Kotb Elmahallawy, E. Molecular characterization of virulence and antibiotic resistance genes among *Salmonella* serovars isolated from broilers in Egypt. *Lett. Appl. Microbiol.* **2019**, *68*, 188–195. [CrossRef] [PubMed]
78. Kaniga, K.; Trollinger, D.; Galán, J.E. Identification of two targets of the type III protein secretion system encoded by the inv and spa loci of *Salmonella* Typhimurium that have homology to the *Shigella IpaD* and *IpaA* proteins. *J. Bacteriol.* **1995**, *177*, 7078–7085. [CrossRef] [PubMed]
79. Chen, L.M.; Kaniga, K.; Galán, J.E. *Salmonella* spp. are cytotoxic for cultured macrophages. *Mol. Microbiol.* **1996**, *21*, 1101–1115. [CrossRef] [PubMed]
80. Hensel, M.; Shea, J.E.; Waterman, S.R.; Mundy, R.; Nikolaus, T.; Banks, G.; Vazquez-Torres, A.; Gleeson, C.; Fang, F.C.; Holden, D.W. Genes encoding putative effector proteins of the type III secretion system of *Salmonella* pathogenicity island 2 are required for bacterial virulence and proliferation in macrophages. *Mol. Microbiol.* **1998**, *30*, 163–174. [CrossRef] [PubMed]
81. Cirillo, D.M.; Valdivia, R.H.; Monack, D.M.; Falkow, S. Macrophage-dependent induction of the *Salmonella* pathogenicity island 2 type III secretion system and its role in intracellular survival. *Mol. Microbiol.* **1998**, *30*, 175–188. [CrossRef] [PubMed]
82. Galyov, E.E.; Wood, M.W.; Rosqvist, R.; Mullan, P.B.; Watson, P.R.; Hedges, S.; Wallis, T.S. A secreted effector protein of *Salmonella* Dublin is translocated into eukaryotic cells and mediates inflammation and fluid secretion in infected ileal mucosa. *Mol. Microbiol.* **1997**, *25*, 903–912. [CrossRef]
83. Gibson, D.L.; White, A.P.; Rajotte, C.M.; Kay, W.W. AgfC and AgfE facilitate extracellular thin aggregative fimbriae synthesis in *Salmonella* Enteritidis. *Microbiology* **2007**, *153*, 1131–1140. [CrossRef]
84. Yoo, A.Y.; Yu, J.E.; Yoo, H.; Lee, T.H.; Lee, W.H.; Oh, J.I.; Kang, H.Y. Role of sigma factor E in regulation of *Salmonella Agf* expression. *Biochem. Biophys. Res. Commun.* **2013**, *430*, 131–136. [CrossRef]
85. Collinson, S.K.; Doig, P.C.; Doran, J.L.; Clouthier, S.; Trust, T.J.; Kay, W.W. Thin, aggregative fimbriae mediate binding of *Salmonella* Enteritidis to fibronectin. *J. Bacteriol.* **1993**, *175*, 12–18. [CrossRef]
86. Borsoi, A.; Santin, E.; Santos, L.R.; Salle, C.T.; Moraes, H.L.; Nascimento, V.P. Inoculation of newly hatched broiler chicks with two Brazilian isolates of *Salmonella* Heidelberg strains with different virulence gene profiles, antimicrobial resistance, and pulsed field gel electrophoresis patterns to intestinal changes evaluation. *Poult. Sci.* **2009**, *88*, 750–758. [CrossRef] [PubMed]
87. Den Bakker, H.C.; Moreno Switt, A.I.; Govoni, G.; Cummings, C.A.; Ranieri, M.L.; Degoricija, L.; Hoelzer, K.; Rodriguez-Rivera, L.D.; Brown, S.; Bolchacova, E.; et al. Genome sequencing reveals diversification of virulence factor content and possible host adaptation in distinct subpopulations of *Salmonella enterica*. *BMC Genomics* **2011**, *12*, 425. [CrossRef] [PubMed]
88. Siddiky, N.A.; Sarker, M.S.; Khan, M.S.R.; Begum, R.; Kabir, M.E.; Karim, M.R.; Rahman, M.T.; Mahmud, A.; Samad, M.A. Virulence and antimicrobial resistance profiles of *Salmonella enterica* serovars isolated from chicken at wet markets in Dhaka, Bangladesh. *Microorganisms* **2021**, *9*, 952. [CrossRef] [PubMed]
89. Campioni, F.; Moratto Bergamini, A.M.; Falcão, J.P. Genetic diversity, virulence genes and antimicrobial resistance of *Salmonella* Enteritidis isolated from food and humans over a 24-year period in Brazil. *Food Microbiol.* **2012**, *32*, 254–264. [CrossRef] [PubMed]
90. Amini, K.; Salehi, T.; Nikbakht Brujeni, G.; Ranjbar, R.; Amini, J.; Ashrafganjooei, S. Molecular detection of invA and spv virulence genes in *Salmonella* Enteritidis isolated from human and animals in Iran. *Afr. J. Microbiol. Res.* **2010**, *4*, 2202–2210.
91. Huehn, S.; La Ragione, R.M.; Anjum, M.; Saunders, M.; Woodward, M.J.; Bunge, C.; Helmuth, R.; Hauser, E.; Guerra, B.; Beutlich, J.; et al. Virulotyping and antimicrobial resistance typing of *Salmonella enterica* serovars relevant to human health in Europe. *Foodborne Pathog. Dis.* **2010**, *7*, 523–535. [CrossRef] [PubMed]
92. Aslam, M.; Checkley, S.; Avery, B.; Chalmers, G.; Bohaychuk, V.; Gensler, G.; Reid-Smith, R.; Boerlin, P. Phenotypic and genetic characterization of antimicrobial resistance in *Salmonella* serovars isolated from retail meats in Alberta, Canada. *Food Microbiol.* **2012**, *32*, 110–117. [CrossRef] [PubMed]
93. O'Regan, E.; Quinn, T.; Frye, J.G.; Pages, J.M.; Porwollik, S.; Fedorka-Cray, P.J.; McClelland, M.; Fanning, S. Fitness costs and stability of a high-level ciprofloxacin resistance phenotype in *Salmonella enterica* serotype Enteritidis: Reduced infectivity associated with decreased expression of *Salmonella* pathogenicity island 1 genes. *Antimicrob. Agents Chemother.* **2010**, *54*, 367–374. [CrossRef]
94. Gunn, J.S.; Ryan, S.S.; Van Velkinburgh, J.C.; Ernst, R.K.; Miller, S.I. Genetic and functional analysis of a *PmrA-PmrB*-regulated locus necessary for lipopolysaccharide modification, antimicrobial peptide resistance, and oral virulence of *Salmonella enterica* serovar Typhimurium. *Infect. Immun.* **2000**, *68*, 6139–6146. [CrossRef]

95. Tamayo, R.; Ryan, S.S.; McCoy, A.J.; Gunn, J.S. Identification and genetic characterization of *PmrA*-regulated genes and genes involved in polymyxin B resistance in *Salmonella enterica* serovar Typhimurium. *Infect. Immun.* **2002**, *70*, 6770–6778. [CrossRef]
96. Eswarappa, S.M.; Panguluri, K.K.; Hensel, M.; Chakravortty, D. The yejABEF operon of *Salmonella* confers resistance to antimicrobial peptides and contributes to its virulence. *Microbiology* **2008**, *154*, 666–678. [CrossRef] [PubMed]
97. Nilsson, A.I.; Zorzet, A.; Kanth, A.; Dahlström, S.; Berg, O.G.; Andersson, D.I. Reducing the fitness cost of antibiotic resistance by amplification of initiator tRNA genes. *Proc. Natl. Acad. Sci. USA* **2006**, *103*, 6976–6981. [CrossRef] [PubMed]
98. Andersson, D.I.; Hughes, D. Antibiotic resistance and its cost: Is it possible to reverse resistance? *Nat. Rev. Microbiol.* **2010**, *8*, 260–271. [CrossRef] [PubMed]

microorganisms MDPI

Review

Salmonella enterica Outbreaks Linked to the Consumption of Tahini and Tahini-Based Products

Geneviève Coulombe [1] and Sandeep Tamber [2,*]

1 Microbiology Evaluation Division, Bureau of Microbial Hazards, Food Directorate, Health Canada, 251 Sir Frederick Banting Driveway, Ottawa, ON K1A 0K9, Canada
2 Microbiology Research Division, Bureau of Microbial Hazards, Food Directorate, Health Canada, 251 Sir Frederick Banting Driveway, Ottawa, ON K1A 0K9, Canada
* Correspondence: sandeep.tamber@hc-sc.gc.ca

Abstract: *Salmonella* is a leading cause of bacterial foodborne illness in the world. Although typically associated with foods of animal origin, low-moisture foods, such as tahini, are quickly gaining recognition as an important vehicle of *Salmonella* exposure. This review offers the Canadian perspective on the issue of *Salmonella* in tahini and tahini-based products. A summary of several recent food product recalls and foodborne outbreaks related to the presence of *Salmonella* in tahini and tahini-based products such as halva are presented. The properties of the food vehicles, their production practices, and potential routes of contamination are discussed. Particular focus is placed on the ecology of *Salmonella* in the tahini production continuum, including its survival characteristics and response to intervention technologies.

Keywords: foodborne illness; foodborne outbreak; low-moisture foods; *Salmonella*; *Salmonella* detection; sesame seeds; tahini; halva

Citation: Coulombe, G.; Tamber, S. *Salmonella enterica* Outbreaks Linked to the Consumption of Tahini and Tahini-Based Products. *Microorganisms* **2022**, *10*, 2299. https://doi.org/10.3390/microorganisms10112299

Academic Editor: Abiodun Adewale Adesiyun

Received: 14 October 2022
Accepted: 17 November 2022
Published: 19 November 2022

Publisher's Note: MDPI stays neutral with regard to jurisdictional claims in published maps and institutional affiliations.

1. Introduction

Salmonellae are an important cause of human illnesses worldwide. It is estimated that nontyphoidal *Salmonella* causes 93.8 million illnesses, of which an estimated 80.3 million are foodborne, and 155,000 deaths each year, worldwide [1]. Most salmonellosis cases are sporadic, meaning they are not linked to known outbreaks [2]. Outbreaks linked to *Salmonella* are often associated with the consumption of foods of animal origin or fresh produce [3]. Multiple outbreaks involving other food commodities have also been reported. Particularly problematic are low-moisture foods, which are defined as foods having a water activity (a_W) of 0.85 or below. This category of foods includes chocolate, peanut butter, and tahini, all of which have been linked to multiple large outbreaks of salmonellosis across the globe.

Low-moisture foods are generally regarded as low-risk foods because they cannot support the growth of pathogenic bacteria, including those belonging to the genus *Salmonella. Salmonella*, however, has the capacity to survive in low-moisture foods for extended periods, up to several years. This prolonged survival increases the risk associated with the consumption of low-moisture foods. As shelf-stable products, a contaminated batch of low-moisture food has the potential to be ingested over time on multiple occasions by multiple people, leading to a wider temporal and geographic distribution of the pathogen than with other food commodities, and a consequent higher number of cases. Three of the largest salmonellosis outbreaks reported in the published literature are linked to low-moisture foods; dried cuttlefish snacks, paprika, and peanut butter with 1505, 1000, and 715 reported cases, respectively [4–6].

In 2004, the World Health Organization's International Food Safety Authorities Network (INFOSAN) identified sesame-based foods (i.e., tahini and halva from the Middle Eastern region) as an unusual food source for *Salmonella* contamination [7]. To date, twenty salmonellosis outbreaks linked to the consumption of tahini and tahini-based products (e.g., halva and hummus) have been identified worldwide (Table 1) [8]. Overall, these outbreaks resulted in 1662 salmonellosis cases, 88 hospitalisations, and 1 death.

In Canada, from 2011–2021, tahini and tahini-based products (e.g., halva and hummus) have been subject to multiple recalls and updated recalls due to the presence of *Salmonella* [9]. In Canada, salmonellosis illnesses linked to the consumption of tahini imported from Israel and hummus were identified in 2018 and 2020, respectively (Table 1) [10]. In 2019, eight Canadian cases were identified by whole genome sequencing (WGS) as being part of a multicountry outbreak linked to tahini and halva imported from Syria (Table 1). Indeed, tahini, halva, and other products made from sesame seeds were labelled, by the Public Health Agency of Canada (PHAC), as being in the "high-risk" food category due to their potential of being contaminated by *Salmonella* [11].

This review offers the Canadian perspective on the issue of *Salmonella* in tahini and tahini-based products and provides an overview of the available literature concerning the ecology of *Salmonella* in sesame seeds, tahini, and halva. Intrinsic properties of sesame seeds, tahini, and halva will be summarized, along with extrinsic factors that can lead to contamination along the processing continuum. Features of the pathogen allowing for long-range survival, mitigation strategies, and areas for further research will also be covered.

Table 1. Known salmonellosis outbreaks linked to tahini and tahini-based products, 1995–2022.

	Year	Products	*Salmonella* Serovar	Country	Cases (Hospital/ Deaths)	Country of Origin	References
1	1995	Tahini	Brandenburg	USA	137	NA	[12]
2	2001	Halva	Typhimurium DT104	Australia, New Zealand, Sweden, Norway	62	Turkey	[13–17]
3	2002	Tahini	Montevideo	Australia	55	Egypt	[18]
4	2003	Tahini	Montevideo	New Zealand	10	Lebanon	[18]
5	2003	Tahini	Montevideo	Australia	3	Lebanon	[18,19]
6	2004	Hummus, eggplant dip	Typhimurium DT197	Australia	173 (25)	NA	[8]
7	2007	Hummus	Heidelberg	USA	802 (29)	NA	[20]
8	2007	Hummus	Heidelberg	USA	11	NA	[8]
9	2007	Hummus	Muenchen	UK	6	NA	[21]
10	2010	Hummus	Typhimurium	Australia	45 (8)	NA	[22]
11	2011	Tahini, hummus	Bovismorbificans	USA	23	Lebanon	[23,24]
12	2012	Tahini	Montevideo, Mbandaka, Maasticht	New Zealand	16 (3)	Turkey	[25]
13	2013	Tahini	Montevideo, Mbandaka	USA	16 (1/1)	Turkey	[24,26]
14	2016–2017	Tahini, sesame seeds	Vari (11:z41:e,n,z15)	Germany, Czech Republic, Luxembourg, UK, Greece	47 (12)	Greece	[27]
15	2017	Hummus	Thompson	USA	13	NA	[24]
16	2018	Tahini, hummus	Concord	Israel, USA	45	Israel	[28]
17	2019	Tahini	Concord	USA	6 (1)	Israel	[29]
18	2019	Hummus	*Salmonella*	USA	9	NA	[30]
19	2020	Hummus	*Salmonella*	Canada	45 (9)	NA	[31]
20	2019–2022	Tahini, halva	Mbandaka, Havana, Amsterdam, Orion, Kintambo, Senftenberg	Germany, Sweden, Norway, Netherlands, Canada, USA, New Zealand	138 (24)	Syria	[32,33]

2. Salmonella

The genus *Salmonella* consists of two species, *S. enterica* and *S. bongori*, that have been subtyped into over 1500 serological variants, or serovars. In terms of human illness, multiple serovars of *S. enterica* account for over 99% of human salmonellosis cases. Children under 5 years old, the elderly, and people with compromised immunity are most vulnerable to salmonellosis and more likely to develop severe symptoms [34,35].

Estimates of the infectious dose for *Salmonella* range from 10^0 to 10^{11} cells [36,37]. Host-related factors such as age and immune status account for much of this variability. The virulence properties of individual strains and the composition of the food matrix can also influence the number of bacteria required to produce an infection. There is evidence that food matrices with a combination of high fat and low water activity (a_W) in a food matrix may protect *Salmonella* from the acidic conditions of the stomach, thus increasing the likelihood of illness from consuming low numbers of the microorganism [38,39]. Analyses of outbreak-associated, high-fat, low-moisture foods such as chocolate and potato chips indicate that inocula ranging from 1 to 45 cells can lead to symptomatic infections [4,40–42].

Concentrations of *Salmonella* ranging from <0.03 MPN/g to 0.46 MPN/g have been found in tahini and tahini-based products linked to salmonellosis outbreaks [18,25]. Based on these levels and the infectious doses reported above, the consumption of as little as 2 g of contaminated product has the potential to lead to illness.

3. Outbreaks Due to the Presence of *Salmonella* in Tahini and Tahini-Based Products

The first reported salmonellosis outbreak linked to the consumption of tahini occurred in 1995 in the United States. A total of 137 individuals got sick due to the presence of *Salmonella* ser. Brandenburg (Outbreak #1, Table 1). Following that, in 2001, there was an international outbreak of *Salmonella* ser. Typhimurium DT104 due to consumption of halva, a ready-to-eat confectionary made with tahini. Cases were reported in Europe as well as Australia and New Zealand (Outbreak #2, Table 1). In 2002 and 2003, three outbreaks of *Salmonella* ser. Montevideo linked to the consumption of tahini occurred in Australia and New Zealand. The implicated tahini products were manufactured in Egypt and Lebanon (Outbreaks #3 to 5, Table 1). Of the twenty outbreaks listed in Table 1, country-of-origin information is available for eleven. These countries include Egypt, Greece, Israel, Lebanon, Syria, and Turkey as the source of the tahini and tahini-based products. The source of the sesame seeds, in most cases, remains unknown.

Diverse *Salmonella* serovars have been isolated during the course of these outbreak investigations, including some reported in multiple outbreaks: Concord (two outbreaks), Heidelberg (two outbreaks), Mbandaka (three outbreaks), Montevideo (five outbreaks), and Typhimurium (three outbreaks). In 2016 and 2017, an outbreak of a previously undescribed *Salmonella* serovar, Vari, occurred in five European Union countries (Outbreak #14, Table 1). The investigation into this outbreak indicated that cross-contamination likely occurred at the Greek manufacturing facility during the production of tahini [27].

Some outbreaks have involved multiple serovars. Six serovars were isolated during the investigation of a multicountry outbreak linked to tahini and halva imported from Syria (Outbreak #20 in Table 1). In that outbreak, cases were identified intermittently from 2019 to 2022 in five European countries (i.e., Denmark, Germany, the Netherlands, Norway, and Sweden), as well as Canada (eight cases), the United States (six cases), and New Zealand (three cases). This outbreak illustrates the worldwide distribution of tahini and tahini-based products. The investigation into this outbreak was not able to determine the root cause of *Salmonella* contamination. However, since the implicated products were in sealed packages, it is likely the contamination event occurred prior to packaging and, hence, exportation [32].

In two instances, illnesses linked to tahini and hummus have occurred in Canada. In 2018, Canadian cases were reported, and the source was identified as imported tahini from Israel. This resulted in a recall [10]. In 2020, a localized outbreak linked to hummus involved a restaurant and a food truck from a single region (Outbreak #19, Table 1). There were 45 laboratory-confirmed cases, and 185 cases which were symptomatic but were not laboratory confirmed. This outbreak illustrates the risks associated not only with tahini, but also with foods prepared from it, such as hummus.

4. The Ecology of *Salmonella* in Tahini and Tahini-Based Products

4.1. *Salmonella and Sesame Seeds*

Sesame seeds are cultivated from the sesame plant (*Sesamum indicum*). They are used as cooking and baking ingredients. Their high oil content (approximately 50% of the seed's weight) makes them a valuable source of oil for cooking, cosmetic, and pharmaceutical applications [13]. Over 95% of the world's sesame seed crop is produced in Africa (43%) and Asia (53%), with worldwide production estimated at seven million metric tonnes. Sudan, Myanmar, the United Republic of Tanzania, and India were the largest producers in 2020 [43]. Many countries, including Canada, do not produce sesame seeds and, therefore, rely on imported sesame seeds to produce tahini and tahini-based products [43]. Approximately 70–75% of the sesame seeds used in the United States are imported, with a few southern states producing seeds domestically [44,45].

S. indicum is a flowering annual that grows in areas with an annual rainfall of 625–1100 mm and temperatures of 27 °C and higher [46]. The sesame plant has an extensive root system and favours well-drained fertile soils with a neutral pH. The fruit of a sesame plant is a capsule that contains multiple seeds. Once the seeds are ripe, the capsule splits open and the seeds are released. These shattering cultivars of sesame seeds are largely harvested by hand. Non-shattering variants have been developed and are more amenable to mechanical harvesting [45]. Excess water, rain, and wind can promote shattering and decrease the yield of the seeds.

Contamination of the sesame seed plant can occur at the preharvest stage. *Salmonella* may be present in the soil, irrigation water, or in fertilizer. Droppings from wild animals carrying *Salmonella* are another potential source of the pathogen [47]. After harvest, sesame seeds are dried to an a_w of 0.5 [48]. Drying the seeds is a critical step in the production process. It is carried out in open areas that can be exposed to dust and aerosols [49,50]. Wet conditions can complicate the drying process as the size and shape of seed prevents aeration and may prolong the time the seeds are exposed to atmospheric elements.

Many *Salmonella* serovars have been recovered from sesame seeds and include Typhimurium DT104, Offa, Tennessee, Poona [13], Montevideo, Stanleyville, Tilene [51], Amsterdam, Anatum, Bareilly, Charity, Cubana, Gaminara, Tennessee, Hvittingfoss, Kentucky, *S. enterica* ssp. *diarizonae* [52], Weltevreden, Newport, Mbandaka, Anatum, Senftenberg, Give, Tennessee, 3, 10: b:-, Havana, Kentucky, Bonn, Cerro, Glostrup, Idikan, Llandoff, Pottsdam, Westminister, *S. enterica* ssp. *arizonae*, and *S. bongori* 48:z4,z24:- [53]. This diversity suggests multiple sources of contamination. Some of these serovars are known clinically and have been implicated in previous outbreaks linked to low-moisture foods (e.g., Montevideo, Tennessee, Newport, and Poona) [6,54–56], whereas others are rarely seen in clinical settings and their virulence properties are unknown (e.g., Tilene, Charity, Idikan, and Amsterdam). It is also not known whether these serovars have geographic significance or unique physiologies that allow them to survive in the sesame seed production environment.

The presence of *Salmonella* in sesame seeds varies widely (Table 2). Surveys of sesame seeds from a variety of locations have shown *Salmonella* prevalence ranging from none detected in 526 samples to 27% out of 359 samples using an analytical unit of 25 g [49]. Studies with enumeration data indicated low levels of contamination with concentrations ranging from 0.06 to 4 MPN/100 g [44] and 360 MPN/100 g [48]. No relationship was observed between *Salmonella* presence and counts of total aerobic bacteria, coliforms, and/or *E. coli* [48,52,57,58]. Surveillance studies of sesame seeds for import into the United States over ten years showed a relatively high prevalence in imported seeds (8–11%) compared to seeds that were collected from domestic retail establishments in the USA (no detections). This shows that postprocessing interventions may be able to control the levels of *Salmonella* on sesame seeds [44,52,53]. These studies also demonstrated a high degree of batch-to-batch variation with respect to *Salmonella* presence, indicating the pathogen is not distributed homogenously within the product, and may require larger sample sizes for reliable detection.

Table 2. Prevalence of *Salmonella* in sesame seeds, tahini, and halva.

Product	Location	Year	Number of Positives	Total Samples	Prevalence	Reference
Sesame seed	Germany	2003	2	16	12.5%	[13]
Sesame seed	UK	2007–2008	13	771	1.7%	[59]
Sesame seed	USA	2006–2009	20	177	11.3%	[53]
Sesame seed	USA	2010	23	233	9.9	[44]
Sesame seed	USA	2013–2014	0	526	0	[52]
Sesame seed	USA	2012–2015 [b]	12	155	7.7	[52]
Sesame seed	Mexico	2018–2019	12	100	12%	[48]
Sesame seed	Burkina Faso	2007–2017	95	359	26.5%	[49]
Sesame seed	Burkina Faso	2021 [a]	0	25	0	[58]
Sesame seed	Sicily	2021 [a]	36	3	8.3	[51]
Sesame seed	Portugal	2022 [a]	0	18	0%	[57]
Tahini	Saudi Arabia	1982–1983	2 [c]	10 [c]	20%	[60]
Tahini	Germany	2003	1	12	8.3%	[13]
Tahini	Lebanon	2015–2017	7	42	17%	[61]
Tahini	Canada	2010–2014	9	2315	0.4%	[62–64]
Halva	Germany	2003	8	71	11.3%	[13]
Halva	Turkey	2010–2015	2	204 [d]	1%	[65]
Halva	Turkey	2007–2008	0	120	0	[66]
Halva	Turkey	2004 [a]	0	68 [d]	0	[67]
Halva	Greece	1997 [a]	0	4 [c]	0	[68]

[a] year of publication, year of sampling is not known; [b] collected at import; [c] number of production plants sampled from; [d] 10 g analytical unit.

4.2. *Salmonella* and Tahini

Tahini is the paste produced from ground sesame seeds, which has a high-fat (57–65%) and low-moisture content (<1%) [69]. On average, tahini has a water activity (a_w) of 0.16 and a pH of 5.9 [70,71]. Tahini is considered a ready-to-eat product, which is stored at ambient temperature with a long shelf life (up to two years) [18]. Traditional to Middle Eastern and Mediterranean cuisine, tahini is used as an ingredient in the preparation of many other ready-to-eat products, such as hummus, baba ghanoush, mutabbal, tarator sauce, and various salad dressings, sauces, and dips [50,72]. These foods pose an additional risk for acquiring salmonellosis since their high-moisture content could amplify *Salmonella* levels should it be present in any of the raw ingredients.

Countries that are key global exporters of tahini include Lebanon, Syria, Egypt, Greece, and Israel [73]. Tahini is gaining popularity in North American and European markets [74]. As an example, 6.8% of the respondents who participated in Foodbook, a Canadian population-based telephone survey, reported the consumption of tahini, halva, and other products made from sesame seeds in the last 7 days [75]. However, this may be an underestimation since multi-ingredient foods, such as sandwiches and salads, may contain tahini as an ingredient that consumers are not aware of [11].

Tahini is typically obtained by milling cleaned, dehulled, and roasted sesame seeds [50,70]. The process includes multiple steps: an initial soak step, a dehulling step, a draining and drying step, a thermal treatment step (roasting), followed by grinding [76]. Some manufacturers might pasteurize the finished product, although it is not known how widespread this practice is and what parameters are being used [61,77]. Other variations in the process have been reported [13,61]. The soaking step is of critical importance for process control. It can be carried out in water or salt water for 12 to 24 h. If *Salmonella* is present on the initial lot of seeds, this practice can amplify the level of *Salmonella* by up to 3 logs prior to the roasting and grinding of seeds [76].

The majority of tahini sold for consumption is made from roasted seeds. Raw tahini, made from unroasted or lightly roasted sesame seeds, represents a small percentage of all tahini sold (e.g., 4.1% in Canada) [50,62]. Roasting parameters are variable and expected to differ among producers. Different roasting temperatures and times (110 to 170 °C for 40 to 180 min), as well as heat-treatment processes (steam and dry heat roasting treatments) have been reported [78–80].

Studies have demonstrated that the roasting of sesame seeds can reduce *Salmonella* levels [76,80]. Torlak and colleagues reported that roasting temperatures and times of 100 °C for 60 min, 130 °C for 50 min, and 150 °C for 30 min were sufficient to generate a minimum of a 5-log reduction in the initial levels of *Salmonella* artificially inoculated onto sesame seeds. The authors noted a period of rapid decline within the first 10 min of roasting followed by a lower rate of reduction. This first ten minutes of heating corresponded to a reduction in the seeds' a_w, from 0.98 to 0.14. This suggests that during the roasting process the surviving population of *Salmonella* cells exhibit an increase in heat resistance as the moisture of the seeds decreases [80].

Similarly, Zhang and colleagues inoculated sesame seeds with 8.5 log CFU/g of *Salmonella*, then soaked and dried them to an a_w of 0.90. When these seeds were roasted at 130 °C using an air-forced oven, *Salmonella* populations decreased below the detection limit (1.7 log CFU/g) within 10 min. However, when the sesame seeds had an a_w of 0.45 before roasting at the same temperature, the decline in cell populations took longer with an approximate 5-log reduction after 60 min, meaning that 3.5 log CFU/g remained on the seeds after this roasting process [76]. This article indicates that the lethality of the roasting process to *Salmonella* present on the sesame seeds depends on the starting a_w.

If *Salmonella* survives the roasting process, it is likely to remain viable throughout the shelf life of the product. Therefore, tahini and tahini-based products may become contaminated by *Salmonella* through insufficient or inadequate roasting of contaminated sesame seeds [76,80]. In addition, it is crucial to avoid cross-contamination after the roasting process. *Salmonella* cross-contamination in low-moisture foods has been traced to factors such as poor sanitation practices, poor equipment design, improper maintenance, and poor ingredient control [47,81].

Guidance documents for manufacturers of tahini and tahini-based products describing *Salmonella* control practices are available. These documents include the Code of Hygienic Practice for Low-Moisture Foods, CXC 75-2015 [39]; the *Hazard Analysis and Critical Control Point Generic Models for Some Traditional Foods: a Manual for the Eastern Mediterranean Region* [70]; and the Control of *Salmonella* in Low-Moisture Foods [81]. The Grocery Manufacturers Association (GMA) of the United States listed the following seven control elements against *Salmonella* contamination: prevent ingress or spread of *Salmonella* in the processing facility, enhance the stringency of hygiene practices and controls in the Primary *Salmonella* Control Area, apply hygienic design principles to building and equipment design, prevent or minimize growth of *Salmonella* within the facility, establish a raw materials/ingredients control program, validate control measures to inactivate *Salmonella*, and establish procedures for verification of *Salmonella* controls and corrective actions [81].

Published *Salmonella* prevalence levels in tahini from Canada, Germany, and the Middle East ranged from 0.4% to 20% (Table 2) [13,60,61]. As with sesame seeds, no relationship was observed between *Salmonella* presence and total aerobic mesophile/coliform/*E. coli* counts. A targeted survey performed in Canada from 2010 to 2014 found a *Salmonella* prevalence of 0.4% (out of 2,315 samples) in tahini. All positive samples were from tahini imported from Middle Eastern countries (i.e., Lebanon, Syria, and Israel), even though almost half of the products tested were domestically produced [62–64]. It is not clear whether the difference between imported and domestic products was due to the origin of the raw ingredients, or to differences in production practices. A study by Alaouie and colleagues comparing tahini made from traditional or more automated methods did not find a difference in *Salmonella* prevalence among the sampled products [61]. Serovars found in tahini include Amsterdam, Havana, Montevideo, Senftenberg [62–64], Typhimurium DT 104 [13], Hadar, Agona, Einsbuettel, and Ubrecht [60].

4.3. *Salmonella and Halva*

Halva is a confectionary widely consumed in the Middle East and Mediterranean. It is primarily a mixture of tahini and sugar with the following specifications: $\geq$24% fat, $\geq$8.5% protein, $\leq$55% sucrose, $\leq$2% fibre, and $\leq$3% water [68]. An analysis of halva produced in a Greek facility indicated an a_W of 0.18 and pH of 6 [68]. Halva is made by mixing tahini with a heated, acidified sugar syrup. The syrup contains a high concentration of glucose, citric or tartaric acid, and soapwort root extract (*Saponaria officinalis*). The syrup is heated to 120–140 °C prior to mixing with the tahini [13,68]. Nuts, cocoa, and other flavourings can be added before portioning and packaging [65]. These additives, particularly cocoa and pistachio, can act as vehicles for the introduction of *Salmonella* into the product. As an illustration, halva products containing cocoa and/or pistachio demonstrated increased *Enterobacteriaceae*/coliform/*E. coli* counts compared to plain halva [65,67]. There was no relationship between the levels of *Enterobacteriaceae*/coliform/*E. coli* and *Salmonella* presence, suggesting these counts can be used as indicators of overall process hygiene as opposed to indicators of pathogen presence. Reported *Salmonella* prevalences in halva ranged from no detections to 11.3%, with Typhimurium DT 104, Poona, and a monophasic strain from serogroup B isolated (Table 2) [13].

5. Survival of *Salmonella* in Sesame Seeds, Tahini, and Halva

5.1. *Survival Studies*

Water activity (a_W) is a key determinant for microbial growth as it is a measure of the amount of water available to carry out essential metabolic functions. *Salmonella* requires a minimal a_W of 0.93 to actively grow and reproduce [36]. Therefore, it cannot grow in sesame seeds, tahini, or halva, but it can survive in these matrices for extended periods [47,82]. Studies on the survival of *Salmonella* in sesame seeds, tahini, and halva have shown survival periods ranging from at least 4 to 12 months [48,68,76,80,83,84]. These studies highlight the public health risk associated with the presence of *Salmonella* in these foods. Even after prolonged storage, the pathogen remains viable, and therefore, capable of causing disease.

The manner in which tahini and halva are stored appear to have an effect on *Salmonella* survival. Increased survival was noted when tahini and halva were stored at refrigeration temperatures as opposed to room temperature [80,84]. It has been proposed that the increased survival at lower temperatures may be linked to changes in the composition of the bacterial membrane that would result in lower rates of permeability [83]. It is also possible that the shift in temperature changes the organization of the lipid–water interface of the tahini/halva in a manner that supports bacterial survival. Work carried out with peanut butter and margarine has shown that the size of the lipid–water droplets can influence microbial survival, with finer droplets presenting more of a hurdle [85,86]. Whether this relationship applies to tahini and halva remains to be investigated.

Other factors that can affect survival include additives such as acids or essential oils derived from spices, both of which led to marginal reductions in *Salmonella* in tahini stored at 10, 25, or 37 °C for 1 month [72,87]. Product packaging can also affect survival, as higher levels of *Salmonella* were recovered from vacuum-sealed halva after eight months of storage compared to air-sealed packages [68].

5.2. Survival Strategies

Salmonella employs a number of adaptations to survive in low-moisture environments [47,82]. Upon inoculation in peanut oil, cells enter a partially dormant state where less than 5% of their genome is transcribed (compared to 78% of cellular genomes cultured in Luria broth) [88]. Among the genes that are transcribed in peanut oil are stress response genes as well as the global regulators sigma D and sigma E, suggesting that in this state, persistence and survival take precedence over growth and metabolism. Upon prolonged storage, *Salmonella* cells can become metabolically dormant where they are viable but not culturable (VBNC) until exposed to favourable growth conditions, at which point they can resuscitate and resume metabolic and other accessory functions such as pathogenesis.

Changes to the cell structure and cell surface have been reported in low-a_w conditions. *Salmonella* cells grown in high NaCl, sucrose, or glycerol form long filaments due to an inhibition of cell septation. Similarly, high osmolarity conditions lead to an alteration of the OmpC/OmpF porin ratios in the outer membrane that lead to altered permeability rates. Permeability can also be affected through increased synthesis of unsaturated fatty acids via FabA [89]. Osmotic pressure is maintained through the upregulation of genes encoding osmoprotectant synthesis and transport proteins (ProP, ProU, and OsmU), as well as increased potassium influx through the Kdp transporter [88]. However, it is not clear if these adaptations extend to sesame seeds, tahini, and halva since the method of desiccation can influence the nature of the bacterial response [82].

Salmonella cells that are tolerant to desiccation undergo a number of physiological changes that may render them less susceptible to other food control measures such as heating, acidification, and/or treatment with biocides [47]. For example, *Salmonella* cells inoculated onto sesame seeds at a starting a_w of 0.45 exhibited a slower rate of decline compared to cells on seeds with an $a_w \geq 0.95$ when the seeds were roasted at 130 °C [76].

There may be a relationship between biofilm formation and desiccation resistance, since cells that form biofilms have increased resistance to many ecological stresses [90]. Cells that overproduce the biofilm extracellular matrix components curli and cellulose have an increased tolerance to desiccation stress [91]. Similarly, mutants deficient in the production of an extracellular matrix have reduced desiccation tolerance [92]. However, it is not clear if this relationship holds in high-fat, low-a_w food systems and is an area that requires further study.

The ability of *Salmonella* to tolerate desiccation varies from strain to strain. Norberto and colleagues reported in a study of 37 *Salmonella* strains belonging to 16 serovars log reductions ranging from 0.6 to 2.7 when inoculated in dry soybean meal and stored at 25 °C for 18 h [93]. There were considerable differences within strains of the same serovar; for example, the respective range of log reductions was 0.9, 1.1, and 1.3 for *S. enterica* serovars Ohio, Montevideo, and Havana, respectively [93]. The nature of these differences is not clear, but may reflect differences in gene presence or gene expression that can impact the response to desiccation.

6. Control of *Salmonella* in Sesame Seeds, Tahini, and Tahini-Based Products

Both thermal and nonthermal treatments of low-moisture foods or their raw materials can be used to control the presence of *Salmonella*. In recent years, there has been intensive research efforts into postproduction lethality treatments of tahini and tahini-based products. The intent is to eliminate *Salmonella* without compromising the quality properties of the food products. Some of the inactivation treatments tested on sesame seeds, tahini, or halva are presented below.

6.1. Thermal Treatment

Roasting of the sesame seeds is a key step to control *Salmonella* during the manufacturing of tahini, as described in detail in Section 4.2.

Some manufacturers may also include a thermal treatment applied to the tahini as a final step [77]. Szpinak and colleagues evaluated the efficacy of a final heat treatment by testing the survival of *Salmonella* ser. Typhimurium in tahini treated at temperatures of 70 °C, 80 °C, and 90 °C. Regardless of the test conditions, the *Salmonella* cells had a high rate of survival with a maximal reduction of 3 log CFU/mL after one hour at 90 °C. The inactivation profiles demonstrated a rapid decline in the first two minutes of treatment, followed by a tailing curve at all tested temperatures, suggesting the applied thermal treatments were only able to inactivate a subset of the *Salmonella* population [77].

6.2. Natural Antimicrobials

Some studies evaluated the impact of incorporating natural antimicrobials into tahini in order to control *Salmonella* [72,87,94]. Ten plant essential oils were tested for their antimicrobial activity against four different *Salmonella* serotypes (Typhimurium, Aberdeen, Cubana, and Paratyphi A) using a disc diffusion assay method [87]. In that study, thyme oil and cinnamon oil showed the highest antimicrobial activity. When added to tahini, 2.0% cinnamon oil reduced the numbers of *Salmonella* between 2.4 and 2.8 log CFU/mL, whereas 2.0% thyme oil led to a reduction between 2.2 and 3.3 log CFU/mL by 28 days, depending on the storage temperature (10 °C, 25 °C, or 37 °C). However, the addition of essential oil extracts reduced the quality and consumer acceptability of the tahini, as assessed by a sensory panel [87]. Another study investigated the use of 3% oregano oil in tahini, which led to a reduction of 1.4 and 0.80 log CFU/g after 7 days at 25 °C and 4 °C, respectively, compared to the control [94].

The antibacterial effects of organic acids, such as acetic and citric acids, have also been tested in tahini. The addition of acetic and citric acids at a concentration of 0.5% reduced *Salmonella* ser. Typhimurium by 2.7–4.8 log CFU/mL and 2.5–3.8 log CFU/mL, respectively, in tahini after 28 days at tested temperatures of 10 °C, 21 °C, and 37 °C [72]. Similarly, Xu and colleagues confirmed a reduction in *Salmonella* levels with the use of 0.5% citric acid (0.8 log CFU/g reduction after 7 days at 4 °C) [94].

Antimicrobial substances, such as metals, chemicals, essential oils, enzymes, and bacteriocins, can also be used in packaging materials [95]. More research is needed to evaluate their potential effect on *Salmonella* spp. in tahini and tahini-based products.

6.3. Irradiation

Ionizing radiation can effectively eliminate microbes that cause foodborne illness, including *Salmonella*. Three types of radiation can be used on bulk or packaged products: gamma rays, X-rays, and electron beams [96]. D'Oca and colleagues evaluated the inactivation of *Salmonella* ser. Montevideo by gamma irradiation of sesame seeds. Their results showed that the use of 5 kGy was able to reduce *Salmonella* counts by 5.53 log CFU/g on sesame seeds [51]. The inactivation of *Salmonella* using gamma irradiation was also tested in tahini [97] and halva [98]. In halva, an irradiation dose of 4.0 kGy resulted in a reduction of 2.1 log CFU/g of *Salmonella*, whereas in tahini, an irradiation dose of 2.0 kGy led to a reduction of 4.7 log CFU/g of *Salmonella* [97,98]. In tahini, the irradiation dose of 2.0 kGy did not affect the physical and chemical properties of the product (i.e., color, peroxide, p-anisidine, and acid values) [97].

6.4. Energy-Based Technologies

Two recent studies tested energy-based technologies on sesame seeds and tahini [99,100]. Xu and colleagues tested the use of radiofrequency heating, in combination with cinnamon oil vapor, on sesame seeds. A radio frequency heating at 80 °C for 5 min combined with 0.83 μL/mL of cinnamon oil vapor achieved a reduction of greater than 5 log CFU/g of *Salmonella* ser. Montevideo on sesame seeds. No synergistic effect of the combined treatments was observed, nor was there any reduction in the quality of the seeds [100].

The efficiency of 2450 MHz microwave heating at 220, 330, 440, 550, and 660 W for the inactivation of *Salmonella* was assessed in tahini. *Salmonella* reductions ranging from 3.6 to 4.7 log CFU/g were observed after treatment. After two minutes of treatment, the core temperatures of the tahini samples reached, 45 °C, 70 °C, 103 °C, 119 °C, and 144 °C at

220, 330, 440, 550, and 660 W, respectively. The authors noted that microwave heating did not affect acid, peroxide, p-anisidine, or color values of the tahini up to 90 °C [99].

6.5. Fumigation and Gas Treatment

Fumigants, such as ethylene oxide and propylene oxide, have shown to be effective for achieving significant reductions in microbial populations in low-moisture foods. More precisely, ethylene oxide is used to treat spices, whereas propylene oxide is used with a variety of foods such as nuts, spices, cocoa beans, and dried fruits [101].

In some countries, including the European Union, the use of ethylene oxide in foods is not permitted. In recent years, multiple recalls of sesame seed products due to the presence of ethylene oxide residues occurred in the European Union. It is believed that ethylene oxide was used for fumigating sesame seeds to eradicate *Salmonella* contamination [102].

Golden and colleagues tested the use of chlorine dioxide (ClO_2) to reduce *Salmonella* levels on sesame seeds. A treatment with 500 mg ClO_2/kg reduced the number of *Salmonella* from 7.6 log CFU/g to 4.9 log CFU/g on sesame seeds [103].

7. Monitoring

There are many challenges associated with the detection of *Salmonella* in sesame seeds, tahini, and tahini-based products. The low numbers of pathogen and heterogeneous distribution patterns necessitates large sample sizes, often beyond the capacity of many food-testing laboratories [104]. Further, the physiological state of *Salmonella* in the low-moisture food matrix requires consideration to prevent false-negative results. As an adaptation to the low-moisture environment, *Salmonella* cells may become injured, enter a period of dormancy (such as a viable but not culturable state), or adopt a filamentous shape [105,106]. Primary enrichment in a nonselective broth encourages cells to recover from these states and multiply to detectable levels. Approaches such as the addition of protectants or applying slower rates of rehydration to the primary enrichment may be employed to prevent cytolysis and increase the level of viable *Salmonella* [107,108]. Rapid methods that forgo secondary enrichment and/or plating in favour of molecular methods to detect *Salmonella* such as PCR should be used in conjunction with a primary enrichment step [104]. The requirement for primary enrichment often delays the time to result, and further research is required to understand how the enrichment period can be optimized for low-moisture foods. Given the issues with product testing, process and environmental monitoring are essential to ensure the hygiene of the tahini manufacturing process [109].

8. Areas for Further Research

In the course of preparing this review, the following knowledge gaps concerning *Salmonella* and tahini were identified. Research directed in these areas can provide information required to mitigate the public health risks associated with the presence of *Salmonella* in tahini and tahini-based products:

- Understanding the ecology and potential virulence of *Salmonella* serovars associated with sesame seeds, tahini, and tahini-based products;
- Knowledge of the response of sesame and tahini-associated serovars, and whether findings from different serovars can be generalized to the genus;
- Survival of *Salmonella* in tahini and other colloidal foods, particularly the impact of rheological and physical properties (i.e., particle size);
- *Salmonella* adaptations to high-fat, low-moisture foods such as tahini (e.g., biofilms, filaments, viable but not culturable) and the potential impact on detection and lethality treatments;
- The impact of interventions, alone and in combination, throughout the tahini manufacturing process, particularly at the seed soaking and roasting steps;
- *Salmonella* response in commercial settings to typical sesame seed roasting parameters in order to understand the physiology of cells that survive the roasting process;

- Improved traceability and environmental/process monitoring of sesame seeds, tahini, and tahini-based products to allow for a more detailed root cause analysis of food safety incidents;
- *Salmonella* growth dynamics during primary enrichment to improve detection and enumeration methods, with an emphasis on reducing the time to result.

9. Conclusions

The issue of *Salmonella* in tahini and tahini-based products is one of global concern. The tahini production continuum from sesame seed to consumption is one that transcends multiple borders, and requires international cooperation for effective management, trace-back, and source attribution. Several features of the *Salmonella*–tahini relationship make it an intractable food–pathogen combination as demonstrated by the spate of recent food safety incidents associated with it. The long shelf life and the ready-to-eat nature of tahini make it an ideal vehicle for a long and sustained exposure to *Salmonella*. Further, the high fat, low-moisture environment of tahini improves *Salmonella* survival and resiliency to a variety of lethality treatments. Given the low infectious dose of *Salmonella*, complete elimination from a batch of contaminated tahini would be required to prevent illness, a goal that is difficult to achieve in this food matrix.

The difficulty in eliminating *Salmonella* to safe levels in tahini underscores the importance of preventing contamination. For many tahini producers, roasting is the only lethal step applied in the manufacturing process. A validated roasting procedure with strict ingredient and parameter control is key. Similarly, given the high potential for contamination after roasting, process verification and environmental monitoring of the production facility is important. Correlations between total aerobic mesophile/*Enterobacteriaceae*/coliform/*E. coli* counts and *Salmonella* presence have not been observed, but enumeration of the former groups can provide information on the hygiene of the production process. Verification and monitoring of the entire production continuum, from sesame seed production to tahini distribution, may not be possible given the reliance on imported products and the complexity of supply chains. This complexity also complicates food safety investigations and root cause analyses in the event of food safety incidents. Therefore, it is important for the implicated countries to adhere to a set of internationally agreed-upon standards and procedures to ensure the safe manufacture and consumption of this global food product.

Author Contributions: G.C. and S.T.: conceptualization; writing—original draft preparation, review, and editing. All authors have read and agreed to the published version of the manuscript.

Funding: This research received no external funding.

Acknowledgments: The authors would like to thank Alex Gill and Angela Catford for their critical review of the manuscript and Mariola Mascarenhas (Public Health Agency of Canada) for her assistance with compiling the data included in Table 1.

Conflicts of Interest: The authors declare no conflict of interest.

References

1. Majowicz, S.E.; Musto, J.; Scallan, E.; Angulo, F.J.; Kirk, M.; O'Brien, S.J.; Jones, T.F.; Fazil, A.; Hoekstra, R.M.; International Collaboration on Enteric Disease "Burden of Illness" Studies. The Global Burden of Nontyphoidal *Salmonella* Gastroenteritis. *Clin. Infect. Dis.* **2010**, *50*, 882–889. [CrossRef] [PubMed]
2. WHO. Salmonella (Non-Typhoidal). Available online: https://www.who.int/news-room/fact-sheets/detail/salmonella-(non-typhoidal) (accessed on 13 October 2022).
3. Painter, J.A.; Hoekstra, R.M.; Ayers, T.; Tauxe, R.V.; Braden, C.R.; Angulo, F.J.; Griffin, P.M. Attribution of Foodborne Illnesses, Hospitalizations, and Deaths to Food Commodities by using Outbreak Data, United States, 1998–2008. *Emerg. Infect. Dis.* **2013**, *19*, 407–415. [CrossRef] [PubMed]
4. Lehmacher, A.; Bockemühl, J.; Aleksic, S. Nationwide outbreak of human salmonellosis in Germany due to contaminated paprika and paprika-powdered potato chips. *Epidemiol. Infect.* **1995**, *115*, 501–511. [CrossRef] [PubMed]
5. Niizuma, T.; Terada, K.; Matsuda, K.; Ogita, S.; Kataoka, N. Intrafamilial transmission of *Salmonella oranienburg*. *Pediatr. Int.* **2002**, *44*, 391–393. [CrossRef]

6. Sheth, A.N.; Hoekstra, M.; Patel, N.; Ewald, G.; Lord, C.; Clarke, C.; Villamil, E.; Niksich, K.; Bopp, C.; Nguyen, T.-A.; et al. A National Outbreak of *Salmonella* Serotype Tennessee Infections From Contaminated Peanut Butter: A New Food Vehicle for Salmonellosis in the United States. *Clin. Infect. Dis.* **2011**, *53*, 356–362. [CrossRef] [PubMed]
7. International Food Safety Authorities Network. *Unusual sources of Salmonella*; International Food Safety Authorities Network: Rome, Italy, 2004.
8. Public Health Agency of Canada; (Ottawa, ON, Canada). Personal Communication., 2022.
9. Canadian Food Inspection, Agency. Recalls and Safety Alerts. Available online: https://recalls-rappels.canada.ca/en (accessed on 13 October 2022).
10. Canadian Food Inspection, Agency. Food Recall Warning—Achva and S&F Brand Tahini Products Recalled due to Salmonella. Available online: https://inspection.canada.ca/food-recall-warnings-and-allergy-alerts/2018-11-28/eng/1543459555660/1543459556655 (accessed on 13 October 2022).
11. Tooby, M.; Morton, V.; Nesbitt, A.; Ciampa, N.; Thomas, M.K. Consumption of High-Risk Foods in the Canadian Population, Foodbook Study, 2014 to 2015. *J. Food Prot.* **2021**, *84*, 1925–1936. [CrossRef] [PubMed]
12. Centers for Disease, Control and Prevention. 1995 Foodborne Outbreak Line Listing. Available online: https://www.cdc.gov/foodsafety/pdfs/1995_linelist.pdf (accessed on 13 October 2022).
13. Brockmann, S. International outbreak of Salmonella Typhimurium DT104 due to contaminated sesame seed products—Update from Germany (Baden-Württemberg). *Wkly. Releases (1997–2007)* **2001**, *5*, 1699. [CrossRef]
14. Aavitsland, P.; Alvseike, O.; Guérin, P.J.; Stavnes, T.L. International outbreak of Salmonella Typhimurium DT104—Update from Norway. *Wkly. Releases (1997–2007)* **2001**, *5*, 1701. [CrossRef]
15. De Jong, B.; Andersson, Y.; Fisher, I.S.; A O'Grady, K.; Powling, J. International outbreak of Salmonella Typhimurium DT104—Update from Enter-net. *Wkly. Releases (1997–2007)* **2001**, *5*, 1705. [CrossRef]
16. De Jong, B.; Andersson, Y.; Giesecke, J.; Hellström, L.; Stamer, U.; Wollin, R. Salmonella Typhimurium outbreak in Sweden from contaminated jars of helva (or halva). *Wkly. Releases (1997–2007)* **2001**, *5*, 1715. [CrossRef]
17. Little, C. International outbreak of Salmonella Typhimurium DT104—Update from the United Kingdom. *Wkly. Releases (1997–2007)* **2001**, *5*, 1700. [CrossRef]
18. Unicomb, L.E.; Simmons, G.; Merritt, T.; Gregory, J.; Nicol, C.; Jelfs, P.; Kirk, M.; Tan, A.; Thomson, R.; Adamopoulos, J.; et al. Sesame seed products contaminated with Salmonella: Three outbreaks associated with tahini. *Epidemiol. Infect.* **2005**, *133*, 1065–1072. [CrossRef] [PubMed]
19. OzFoodNet Working, Group. Foodborne Disease Investigation across Australia: Annual Report of the Ozfoodnet Network, 2003. *Commun. Dis. Intell. Q. Rep.* **2004**, *28*, 359–389.
20. Centers for Disease, Control and Prevention. Surveillance for Foodborne Disease Outbreaks—United States, 2007. *Morb. Mortal. Wkly. Rep.* **2010**, *59*, 973–979.
21. Harker, K.S.; Lane, C.; Gormley, F.J.; Adak, G.K. National Outbreaks of *Salmonella* Infection in the UK, 2000–2011. *Epidemiol. Infect.* **2014**, *142*, 601–607. [CrossRef] [PubMed]
22. Torres, M.I.; Lewis, P.; Cook, L.; Cook, P.; Kardamanidis, K.; Shadbolt, C.; Campbell, B. An outbreak of Salmonella Typhimurium linked to a kebab takeaway shop. *Commun. Dis. Intell. Q. Rep.* **2012**, *36*, 101–106. [PubMed]
23. Centers for Disease Control and Prevention. Multistate Outbreak Of salmonella serotype Bovismorbificans Infections Associated with Hummus and Tahini—United States, 2011. *Morb. Mortal. Wkly. Rep.* **2012**, *61*, 944–947.
24. Centers for Disease Control and Prevention. National Outbreak Reporting System (NORS). Available online: https://wwwn.cdc.gov/norsdashboard/ (accessed on 13 October 2022).
25. Paine, S.; Thornley, C.; Wilson, M.; Dufour, M.; Sexton, K.; Miller, J.; King, G.; Bell, S.; Bandaranayake, D.; Mackereth, G. An Outbreak of Multiple Serotypes of *Salmonella* in New Zealand Linked to Consumption of Contaminated Tahini Imported from Turkey. *Foodborne Pathog. Dis.* **2014**, *11*, 887–892. [CrossRef] [PubMed]
26. Centers for Disease Control and Prevention. Salmonella Mbandaka Infections Linked to Tahini Sesame Paste. 2013. Available online: https://www.cdc.gov/salmonella/montevideo-tahini-05-13/ (accessed on 13 October 2022).
27. European Food Safety Authority and European Centre for Disease Prevention and Control. Multi-Country Outbreak of New Salmonella enterica 11:Z41:E,N,Z15 Infections Associated with Sesame Seeds. Available online: https://www.ecdc.europa.eu/sites/default/files/documents/rapid-outbreak-assessment-salmonella-enterica-Greece-14-Jun-2017.pdf (accessed on 13 October 2022).
28. Centers for Disease Control and Prevention. Salmonella Infections Linked to Tahini from Achdut Ltd. 2018. Available online: https://www.cdc.gov/salmonella/concord-11-18/index.html (accessed on 13 October 2022).
29. Centers for Disease Control and Prevention. Salmonella Infections Linked to Karawan brand Tahini. Available online: https://www.cdc.gov/salmonella/concord-05-19/index.html (accessed on 13 October 2022).
30. Maryland Department of, Health. Maryland Department of Health investigating Cluster of Salmonella Infections. 2019. Available online: https://health.maryland.gov/newsroom/Pages/Maryland-Department-of-Health-investigating-cluster-of-salmonella-infections.aspx (accessed on 13 October 2022).
31. Lambton Public, Health. Salmonellosis Outbreak Investigation Closed Locally. Available online: https://lambtonpublichealth.ca/salmonellosis-outbreak-investigation-closed-locally/ (accessed on 13 October 2022).

32. European Food Safety Authority. Multi-Country Outbreak of Multiple Salmonella enterica Serotypes Linked to Imported Sesame-Based Products. Available online: https://www.efsa.europa.eu/en/supporting/pub/en-6922 (accessed on 13 October 2022).
33. Ministry of Health. Monthly Notifiable Disease Surveillance Report—July 2022. Available online: https://surv.esr.cri.nz/index.php?we_objectID=5232 (accessed on 13 October 2022).
34. Centers for Disease, Control, and Prevention. Salmonella. Available online: https://www.cdc.gov/salmonella/general/index.html (accessed on 13 October 2022).
35. U. S. Food and Drug Administration. Get the Facts about Salmonella. Available online: https://www.fda.gov/animal-veterinary/animal-health-literacy/get-facts-about-salmonella (accessed on 13 October 2022).
36. D'Aoust, J.Y.; Maurer, J. Salmonella Species. In *Food Microbiology: Fundamentals and Frontiers*; Doyle, M.P., Beauchat, L.R., Montville, T.J., Eds.; AMS Press: Washington, DC, USA, 2007; pp. 187–236.
37. United States Food and Drug Administration. Bad Bug Book. Available online: https://www.fda.gov/media/83271/download (accessed on 7 October 2021).
38. Aviles, B.; Klotz, C.; Smith, T.; Williams, R.; Ponder, M. Survival of *Salmonella enterica* Serotype Tennessee during Simulated Gastric Passage Is Improved by Low Water Activity and High Fat Content. *J. Food Prot.* **2013**, *76*, 333–337. [CrossRef] [PubMed]
39. Codex Alimentarius, Commission. Code of Hygienic PRACTICE for low-Moisture Foods CXC 75-2015. Available online: https://www.fao.org/fao-who-codexalimentarius/sh-proxy/en/?lnk=1&url=https%253A%252F%252Fworkspace.fao.org%252Fsites%252Fcodex%252FStandards%252FCXC%2B75-2015%252FCXC_075e.pdf (accessed on 13 October 2022).
40. Greenwood, M.H.; Hooper, W.L. Chocolate bars contaminated with *Salmonella napoli*: An infectivity study. *BMJ* **1983**, *286*, 1394. [CrossRef]
41. Kapperud, G.; Gustavsen, S.; Hellesnes, I.; Hansen, A.H.; Lassen, J.; Hirn, J.; Jahkola, M.; A Montenegro, M.; Helmuth, R. Outbreak of Salmonella typhimurium infection traced to contaminated chocolate and caused by a strain lacking the 60-megadalton virulence plasmid. *J. Clin. Microbiol.* **1990**, *28*, 2597–2601. [CrossRef] [PubMed]
42. Vought, K.J.; Tatini, S.R. *Salmonella enteritidis* Contamination of Ice Cream Associated with a 1994 Multistate Outbreak. *J. Food Prot.* **1998**, *61*, 5–10. [CrossRef] [PubMed]
43. FAOSTAT. Crops and Livestock Products. Available online: https://www.fao.org/faostat/en/#data/QCL (accessed on 29 August 2022).
44. Van Doren, J.M.; Blodgett, R.J.; Pouillot, R.; Westerman, A.; Kleinmeier, D.; Ziobro, G.C.; Ma, Y.; Hammack, T.S.; Gill, V.; Muckenfuss, M.F.; et al. Prevalence, level and distribution of Salmonella in shipments of imported capsicum and sesame seed spice offered for entry to the United States: Observations and modeling results. *Food Microbiol.* **2013**, *36*, 149–160. [CrossRef]
45. Agricultural Marketing Resource Center. Sesame Profile. Available online: https://www.agmrc.org/commodities-products/grains-oilseeds/sesame-profile (accessed on 6 October 2022).
46. Terefe, G.; Adugna, W.; Muez, B.; Hagos, T. Sesame Production Manual; Ethiopian Institute of Agricultural Research Embassy of the Kingdom of the Netherlands: 2012; ISBN 978-99944-53-80-8. Available online: https://www.researchgate.net/publication/301771324_Sesame_production_manual (accessed on 18 November 2022).
47. Podolak, R.; Enache, E.; Stone, W.; Black, D.G.; Elliott, P.H. Sources and Risk Factors for Contamination, Survival, Persistence, and Heat Resistance of Salmonella in Low-Moisture Foods. *J. Food Prot.* **2010**, *73*, 1919–1936. [CrossRef] [PubMed]
48. Juarez Arana, C.D.; Peniche, R.A.M.; Martinez, M.G.; Iturriaga, M.H. Microbiological Profile, Incidence, and Behavior of Salmonella on Seeds Traded in Mexican Markets. *J. Food Prot.* **2021**, *84*, 99–105. [CrossRef]
49. Compaore, M.K.A.; Yougbara, V.M.; Dembala, R.; Nikiama, F.; Elie, K.; Barro, N. Retrospective Study of the Contamination of Exported Sesame by *Salmonella* Species from 2007 to 2017 in Burkina Faso. *Afr. J. Agric. Res.* **2020**, *16*, 1141–1147.
50. Lake, R.; King, N.; Cressey, P.; Gilbert, S. Risk Profile: Salmonella (Non typhoidal) in High Lipid Foods. Available online: https://www.mpi.govt.nz/dmsdocument/24248-Salmonella-non-typhoidal-in-high-lipid-foods-made-from-sesame-seeds-peanuts-or-cocoa-beans (accessed on 13 October 2022).
51. D'Oca, M.C.; di Noto, A.M.; Bartolotta, A.; Parlato, A.; Nicastro, L.; Sciortino, S.; Cardamone, C. Assessment of Contamination of *Salmonella* spp. In Imported Black Pepper and Sesame Seed and Salmonella Inactivation by Gamma Irradiation. *Ital. J. Food Saf.* **2021**, *10*, 8914. [CrossRef]
52. Zhang, G.; Hu, L.; Pouillot, R.; Tatavarthy, A.; Van Doren, J.M.; Kleinmeier, D.; Ziobro, G.C.; Melka, D.; Wang, H.; Brown, E.W.; et al. Prevalence of Salmonella in 11 Spices Offered for Sale from Retail Establishments and in Imported Shipments Offered for Entry to the United States. *J. Food Prot.* **2017**, *80*, 1791–1805. [CrossRef]
53. Van Doren, J.M.; Kleinmeier, D.; Hammack, T.S.; Westerman, A. Prevalence, serotype diversity, and antimicrobial resistance of Salmonella in imported shipments of spice offered for entry to the United States, FY2007–FY2009. *Food Microbiol.* **2013**, *34*, 239–251. [CrossRef] [PubMed]
54. Haendiges, J.; Blessington, T.; Zheng, J.; Davidson, G.; Miller, J.D.; Hoffmann, M. Complete Genome Sequences of Four Salmonella enterica subsp. enterica Serovar Senftenberg and Montevideo Isolates Associated with a 2016 Multistate Outbreak in the United States. *Genome Announc.* **2018**, *6*, e00630-18. [CrossRef]
55. Harvey, R.R.; Marshall, K.E.H.; Burnworth, L.; Hamel, M.; Tataryn, J.; Cutler, J.; Meghnath, K.; Wellman, A.; Irvin, K.; Isaac, L.; et al. International Outbreak of Multiple Salmonella Serotype Infections Linked to Sprouted Chia Seed Powder—USA and Canada, 2013–2014. *Epidemiol. Infect.* **2017**, *145*, 1535–1544. [CrossRef]

56. Jones, G.; de la Gandara, M.P.; Herrera-Leon, L.; Herrera-Leon, S.; Martinez, C.V.; Hureaux-Roy, R.; Abdallah, Y.; Nisavanh A.; Fabre, L.; Renaudat, C.; et al. Outbreak of Salmonella enterica serotype Poona in infants linked to persistent Salmonella contamination in an infant formula manufacturing facility, France, August 2018 to February 2019. *Eurosurveillance* **2019**, *24*, 1900161 [CrossRef]
57. Silva, D.; Nunes, P.; Melo, J.; Quintas, C. Microbial quality of edible seeds commercially available in southern Portugal. *AIMS Microbiol.* **2022**, *8*, 42–52. [CrossRef] [PubMed]
58. Compaoré, M.K.A.; Bazie, B.S.R.; Nikiema, M.E.M.; Dakené, V.M.; Dembélé, R.; Kpoda, D.S.; Kabré, E.; Barro, N. Assessment of the Sanitary Quality of Ready to Eat Sesame, a Low Moisture Street Food from Burkina Faso. *BMC Microbiol.* **2021**, *21*, 207 [CrossRef] [PubMed]
59. Willis, C.; Little, C.L.; Sagoo, S.; de Pinna, E.; Threlfall, J. Assessment of the microbiological safety of edible dried seeds from retail premises in the United Kingdom with a focus on *Salmonella* spp. *Food Microbiol.* **2009**, *26*, 847–852. [CrossRef] [PubMed]
60. Ayaz, M.; Sawaya, W.N.; Al-Sogair, A. Microbiological Quality of Tehineh Manufactured in Saudi Arabia. *J. Food Prot.* **1986**, *49* 504–506. [CrossRef] [PubMed]
61. Alaouie, Z.; Hallal, N.; AlKhatib, A.; Khachfe, H.M. Assessing the Microbial Quality of Tahini (Sesame Paste) in Lebanon. In *Global Health 2017: The Sixth International Conference on Global Health Challenges*; IARIA: Barcelona, Spain, 2017.
62. Canadian Food Inspection Agency. Targeted Survey Investigating Salmonella, Listeria Monocytogenes and Generic E. *coli in Tahini 2012–2014*. Available online: https://inspection.canada.ca/food-safety-for-industry/food-chemistry-and-microbiology/food safety-testing-bulletin-and-reports/salmonella-listeria-monocytogenes-and-generic-e-co/eng/1645637293925/1645637294253 (accessed on 13 October 2022).
63. Canadian Food Inspection Agency. 2010–2011 Salmonella and Generic E. *coli in Tahini*. Available online: https://inspection canada.ca/food-safety-for-industry/food-chemistry-and-microbiology/food-safety-testing-bulletin-and-reports/salmonella and-generic-e-coli-in-tahini/eng/1430919384198/1430919384776 (accessed on 13 October 2022).
64. Canadian Food Inspection Agency. 2011–2012 Salmonella and Generic E. *coli in Tahini and Sesame Seeds*. Available on line: https://inspection.canada.ca/food-safety-for-industry/food-chemistry-and-microbiology/food-safety-testing-bulletin and-reports/salmonella-and-generic-e-coli-in-tahini-and-sesame/eng/1430919620350/1430919620865 (accessed on 13 October 2022).
65. Var, I.; Uzunlu, S.; Mavis, F.G. Microbial Quality of a Confection (Tahini Halva) Manufactured in Turkey. *J. Food Saf. Food Qual* **2020**, *71*, 124–128.
66. Kahraman, T.; Issa, G.; Ozmen, G.; Buyukunal, S. Microbiological and chemical quality of tahini halva. *Br. Food J.* **2010**, *112* 608–616. [CrossRef]
67. Sengun, I.Y.; Hancioglu, O.; Karapinar, M. Microbiological profile of helva sold at retail markets in Izmir city and the survival of Staphylococcus aureus in this product. *Food Control* **2005**, *16*, 840–844. [CrossRef]
68. Kotzekidou, P. Microbial Stability and Fate of *Salmonella* Enteritidis in Halva, a Low-Moisture Confection. *J. Food Prot.* **1998**, *61* 181–185. [CrossRef]
69. Abu-Jdayil, B.; Al-Malah, K.; Asoud, H. Rheological characterization of milled sesame (tehineh). *Food Hydrocoll.* **2002**, *16*, 55–61 [CrossRef]
70. World Health, Organization. Hazard Analysis and Critical Control Point Generic Models for Some Traditional Foods: A manual for the Eastern Mediterranean Region. Available online: https://apps.who.int/iris/handle/10665/119885 (accessed on 13 October 2022).
71. Yamani, M.I.; Isa, J.K. Microbiological Quality of Tehena and Development of a Generic HACCP Plan for Its Production. *World J Agric. Sci.* **2006**, *2*, 290–297.
72. Al-Nabulsi, A.A.; Olaimat, A.N.; Osaili, T.M.; Shaker, R.R.; Elabedeen, N.Z.; Jaradat, Z.W.; Abushelaibi, A.; Holley, R.A. Use of acetic and citric acids to control Salmonella Typhimurium in tahini (sesame paste). *Food Microbiol.* **2014**, *42*, 102–108. [CrossRef] [PubMed]
73. Technavio. Tahini Market by Application and Geography—Forecast and Analysis 2021–2025. Available online: https://www technavio.com/report/tahini-market-industry-analysis (accessed on 13 October 2022).
74. Zion Market Research. Tahini Market—Global Industry Analysis. 2019. Available online: https://www.zionmarketresearch.com/ report/tahini-market (accessed on 13 October 2022).
75. Public Health Agency of Canada. *Foodbook Report*; Public Health Agency of Canada: Ottawa, Canada, 2015; ISBN 978-0-660-03357 0.
76. Zhang, Y.; Keller, S.E.; Grasso-Kelley, E.M. Fate of Salmonella throughout Production and Refrigerated Storage of Tahini. *J. Food Prot.* **2017**, *80*, 940–946. [CrossRef] [PubMed]
77. Szpinak, V.; Ganz, M.; Yaron, S. Factors affecting the thermal resistance of Salmonella Typhimurium in tahini. *Food Res. Int.* **2022** *155*, 111088. [CrossRef] [PubMed]
78. Kahyaoglu, T.; Kaya, S. Modeling of moisture, color and texture changes in sesame seeds during the conventional roasting *J. Food Eng.* **2006**, *75*, 167–177. [CrossRef]
79. El-Adawy, T.A.; Mansour, E.H. Nutritional and Physicochemical Evaluations of Tahina (Sesame Butter) Prepared from Heat Treated Sesame Seeds. *J. Sci. Food Agric.* **2000**, *80*, 2005–2011. [CrossRef]

80. Torlak, E.; Sert, D.; Serin, P. Fate of Salmonella during sesame seeds roasting and storage of tahini. *Int. J. Food Microbiol.* **2013**, *163*, 214–217. [CrossRef] [PubMed]
81. The Association of Food, Beverage and Consumer Products Companies. Control of Salmonella in Low-Moisture Foods. Available online: https://graphics8.nytimes.com/packages/pdf/business/20090515_moss_ingredients/SalmonellaControlGuidance.pdf (accessed on 13 October 2022).
82. Finn, S.; Condell, O.; McClure, P.; Amézquita, A.; Fanning, S. Mechanisms of survival, responses and sources of Salmonella in low-moisture environments. *Front. Microbiol.* **2013**, *4*, 331. [CrossRef]
83. Osaili, T.; Al-Nabulsi, A.; Nazzal, D.; Al-Holy, M.; Olaimat, A.; Obaid, R.; Holley, R. Effect of water activity and storage of tahini on the viability of stressed *Salmonella* serovars. *Food Sci. Technol.* **2021**, *11*, 144–150. [CrossRef]
84. Osaili, T.M.; Al-Nabulsi, A.A.; Nazzal, D.S.; Shaker, R.R. Effect of Storage Temperatures and Stresses on the Survival of Salmonella spp. In Halva. *Lett. Appl. Microbiol.* **2017**, *65*, 403–409. [CrossRef] [PubMed]
85. Burnett, S.L.; Gehm, E.R.; Weissinger, W.R.; Beuchat, L.R. Survival of *Salmonella* in Peanut Butter and Peanut Butter Spread. *J. Appl. Microbiol.* **2000**, *89*, 472–477. [CrossRef] [PubMed]
86. Delamarre, S.; Batt, C. The microbiology and historical safety of margarine. *Food Microbiol.* **1999**, *16*, 327–333. [CrossRef]
87. Al-Nabulsi, A.A.; Osaili, T.M.; Olaimat, A.N.; Almasri, W.E.; Ayyash, M.; Al-Holy, M.A.; Jaradat, Z.W.; Obaid, R.S.; Holley, R.A. Inactivation of *Salmonella* spp. in tahini using plant essential oil extracts. *Food Microbiol.* **2019**, *86*, 103338. [CrossRef] [PubMed]
88. Deng, X.; Li, Z.; Zhang, W. Transcriptome sequencing of *Salmonella enterica* serovar Enteritidis under desiccation and starvation stress in peanut oil. *Food Microbiol.* **2012**, *30*, 311–315. [CrossRef]
89. Chen, W.; Golden, D.; Critzer, F. *Salmonella* survival and differential expression of fatty acid biosynthesis-associated genes in a low-water-activity food. *Lett. Appl. Microbiol.* **2014**, *59*, 133–138. [CrossRef]
90. Marmion, M.; Macori, G.; Whyte, P.; Scannell, A.G.M. Stress response modulation: The key to survival of pathogenic and spoilage bacteria during poultry processing. *Microbiology* **2022**, *168*, 001184. [CrossRef]
91. White, A.P.; Gibson, D.L.; Kim, W.; Kay, W.W.; Surette, M.G. Thin Aggregative Fimbriae and Cellulose Enhance Long-Term Survival and Persistence of *Salmonella*. *J. Bacteriol.* **2006**, *188*, 3219–3227. [CrossRef]
92. Maserati, A.; Fink, R.C.; Lourenco, A.; Julius, M.L.; Diez-Gonzalez, F. General Response of *Salmonella enterica* Serovar Typhimurium to Desiccation: A New Role for the Virulence Factors sopDand sseD in Survival. *PLoS ONE* **2017**, *12*, e0187692. [CrossRef]
93. Norberto, A.P.; Alvarenga, V.O.; Hungaro, H.M.; Sant'Ana, A.S. Desiccation resistance of a large set of Salmonella enterica strains and survival on dry- and wet-inoculated soybean meal through storage. *LWT* **2022**, *158*, 113153. [CrossRef]
94. Xu, Y.; Guan, X.; Lin, B.; Li, R.; Wang, S. Oregano Oil, Epsilon-Polylysine and Citric Acid Assisted Inactivation of *Salmonella* in Two Kinds of Tahini during Thermal Treatment and Storage. *Foods* **2021**, *10*, 1272. [CrossRef] [PubMed]
95. Mauriello, G. Control of Microbial Activity Using Antimicrobial Packaging. In *Antimicrobial Food Packaging*; Barros-Velázquez, J., Ed.; Academic Press: San Diego, CA, USA, 2016; pp. 141–152.
96. Niemara, B.A. Irradiation, Microwave, and Alternative Energy-Based Treatments for Low-Water Activity Foods. In *The Microbiological Safety of Low Water Activity Foods and Spices*; Gurtler, J.B., Doyle, M.P., Kornacki, J.L., Eds.; Springer: New York, NY, USA, 2014; pp. 389–402.
97. Osaili, T.M.; Al-Nabulsi, A.A.; Abubakar, S.A.; Alaboudi, A.R.; Al-Holy, M.A. Feasibility of Using Gamma Irradiation for Inactivation of Starvation-, Heat-, and Cold-Stressed *Salmonella* in Tahini. *J. Food Prot.* **2016**, *79*, 963–969. [CrossRef]
98. Osaili, T.M.; Al-Nabulsi, A.A.; Aljaafreh, T.F. Use of Gamma Radiation for Inactivating *Salmonella* spp., Escherichia coli O157: H7 and Listeria monocytogenes in Tahini Halva. *Int. J. Food Microbiol.* **2018**, *278*, 20–25. [CrossRef] [PubMed]
99. Osaili, T.M.; Al-Nabulsi, A.A.; Al Sheikh, Y.M.; Alaboudi, A.R.; Olaimat, A.N.; Al-Holy, M.; Al-Rousan, W.M.; Holley, R. Inactivation of *Salmonella* spp., *Escherichia coli* O157: H7 and *Listeria monocytogenes* in Tahini by Microwave Heating. *Foods* **2021**, *10*, 2972. [CrossRef] [PubMed]
100. Xu, Y.; Li, R.; Li, K.; Yu, J.; Bai, J.; Wang, S. Inactivation of inoculated *Salmonella* and natural microflora on two kinds of edible seeds by radio frequency heating combined with cinnamon oil vapor. *LWT* **2021**, *154*, 112603. [CrossRef]
101. Anderson, N.M. Recent advances in low moisture food pasteurization. *Curr. Opin. Food Sci.* **2018**, *29*, 109–115. [CrossRef]
102. European Parliamentary Research Service. Recalls of Sesame Seed Products Due to Pesticide Residues. Available online: https://www.europarl.europa.eu/RegData/etudes/ATAG/2021/679077/EPRS_ATA(2021)679077_EN.pdf (accessed on 13 October 2022).
103. Golden, C.E.; Berrang, M.E.; Kerr, W.L.; Harrison, M.A. Slow-release chlorine dioxide gas treatment as a means to reduce *Salmonella* contamination on spices. *Innov. Food Sci. Emerg. Technol.* **2019**, *52*, 256–261. [CrossRef]
104. Bell, R.L.; Jarvis, K.G.; Ottesen, A.R.; McFarland, M.A.; Brown, E.W. Recent and emerging innovations in *Salmonella* detection: A food and environmental perspective. *Microb. Biotechnol.* **2016**, *9*, 279–292. [CrossRef]
105. Ryan, G.; Roof, S.; Post, L.; Wiedmann, M. Evaluation of Rapid Molecular Detection Assays for *Salmonella* in Challenging Food Matrices at Low Inoculation Levels and Using Difficult-to-Detect Strains. *J. Food Prot.* **2015**, *78*, 1632–1641. [CrossRef]
106. Mattick, K.L.; Jørgensen, F.; Legan, J.D.; Cole, M.B.; Porter, J.; Lappin-Scott, H.M.; Humphrey, T.J. Survival and Filamentation of *Salmonella enterica* Serovar Enteritidis PT4 and *Salmonella enterica* Serovar Typhimurium DT104 at Low Water Activity. *Appl. Environ. Microbiol.* **2000**, *66*, 1274–1279. [CrossRef] [PubMed]
107. Enache, E.; Podolak, R.; Kataoka, A.; Harris, L.J. Persistence of Salmonella and Other Bacterial Pathogens in Low-Moisture Foods. In *Control of Salmonella and Other Bacterial Pathogens in Low Moisture Foods*; John Wiley & Sons: Hoboken, NJ, USA, 2017; pp. 67–86.

108. Lang, E.; Zoz, F.; Iaconelli, C.; Guyot, S.; Alvarez-Martin, P.; Beney, L.; Perrier-Cornet, J.-M.; Gervais, P. Recovery Estimation of Dried Foodborne Pathogens Is Directly Related to Rehydration Kinetics. *PLoS ONE* **2016**, *11*, e0160844. [CrossRef] [PubMed]
109. Bourdichon, F.; Betts, R.; Dufour, C.; Fanning, S.; Farber, J.; McClure, P.; Stavropoulou, D.A.; Wemmenhove, E.; Zwietering, M.H.; Winkler, A. Processing Environment Monitoring in Low Moisture Food Production Facilities: Are We Looking for the Right Microorganisms? *Int. J. Food Microbiol.* **2021**, *356*, 109351. [CrossRef] [PubMed]

MDPI
St. Alban-Anlage 66
4052 Basel
Switzerland
www.mdpi.com

Microorganisms Editorial Office
E-mail: microorganisms@mdpi.com
www.mdpi.com/journal/microorganisms

www.ingramcontent.com/pod-product-compliance
Lightning Source LLC
Chambersburg PA
CBHW062056220926
51850CB00010B/2784
9783036588964